Emergent Pollutants in Freshwater Plankton Communities

Emergent Pollutants in Freshwater Plankton Communities introduces the environmental and health monitoring techniques for emergent pollutants and their influences on the community structure of lentic freshwater plankton. It highlights the challenges posed by the improper treatment or disposal of industrial pharmaceutical wastes, which could cause numerous related environmental and health risks. It also suggests possible sustainable mitigation techniques for the treatment of emerging pollutants. Further, it addresses the issues of regulatory and monitoring frameworks and reviews laws governing the management and disposal of wastes. This book will help students, researchers, and professionals address the underlying issues of waste water pollutants from various industries and ideally provide methods to achieve a sustainable and pollutant-free environment for the present and future generations.

- Presents detailed information regarding emergent pollutant effects on freshwater organisms, as well as mitigation and remediation techniques.
- Discusses foundational information regarding issues of wastewater pollutants from pharmaceuticals, personal care products/cosmetics, and other various industries.
- Examines several sustainable mitigation techniques for the treatment of waste pollutants.
- Addresses the issues of regulatory and monitoring frameworks and reviews laws governing the disposal and management of waste.

Emergent Pollutants in Freshwater Plankton Communities

Ecological Effects and Sustainable Mitigation Strategies

Edited by
Osikemekha Anthony Anani and
Maulin P. Shah

CRC Press
Taylor & Francis Group
Boca Raton London New York

CRC Press is an imprint of the
Taylor & Francis Group, an **informa** business

Designed cover image: Shutterstock

First edition published 2025
by CRC Press
2385 NW Executive Center Drive, Suite 320, Boca Raton FL 33431

and by CRC Press
4 Park Square, Milton Park, Abingdon, Oxon, OX14 4RN

CRC Press is an imprint of Taylor & Francis Group, LLC

ISBN: 978-1-032-42481-1 (hbk)
ISBN: 978-1-032-42482-8 (pbk)
ISBN: 978-1-003-36297-5 (ebk)

DOI: 10.1201/9781003362975

Typeset in Times
by Apex CoVantage, LLC

Each editor wishes to dedicate this book to their respective parents, better halves and kids.

Contents

Preface

The natural ecosystems have been contaminated for hundreds of decades by the activities of humans, like industrial waste discharges from pharmaceutical effluents and insecticidal discharges from domestic and agronomic sources. The contamination of the environment could reach a level where the natural living and nonliving components can no longer maintain their innate balances but respond to it in a non-functional state, as in the abiotic components, which will culminate in a pollution status stage or displacement, distortion, disruption, and eventual death of the biotic components, depending on their ability to withstand such physical or biological stressors.

Recently, the release of pollutants into the environment has been linked to the population's demand for a particular product to satisfy their needs and desires. These have led to strains of multifaceted emerging pollutants with known and unknown health and environmental impacts that have affected the global population of humans, plants, and animals, thus reducing their functional and structural diversities, services, and ecosystem processes.

Emergent pollutants or contaminants (EPs and ECs) are natural or synthetic arising chemicals that are not frequently supervised in the ecosystem but pose potential suspected or known health harm to humans and the life forms in the ecosystems (water, air, and land) and their immediate surroundings.

Personal care, cosmetics, and pharmaceutical wastes serve as point sources of EPs and ECs when released in large or small amounts. It has been established by many researchers that EPs and ECs like compounds of perfluoroalkyl, musks, microplastics, triclosan, parabens, and UV filters from cosmetics, personal care products, and pharmaceuticals can persist and bioaccumulate for a long time in the ecosystem, clogging or masking themselves to substances and acting as micropollutants in sewage, polluted or contaminated water, soil, sediment, and air that can also serve as an essential threat to the ecosystem structure and function.

The toxicity of EPs and ECs, when released into the environment in large concentrations, cannot be quantified, and this has provoked global attention to their ecological and human risks. The aquatic biodiversity is not left out of the scourging effects of EPs and ECs. Planktons, which are collective names for organisms like coelenterates, mollusks, crustaceans, protozoans, bacteria, and some algae, represent the class of planktonic-nekton animals commonly and abundantly found in the interfacial and superficial regions of water bodies with low currents. The organisms are special assets in the food chain structure of the aquatic environment and play several roles in ecosystem services as well. However, they are threatened by the influence of emergent pollutants.

The 1992 UN conference on Environment and Development stressed the need to conserve and protect biodiversity from torrential pollution in freshwater habitats and environments supporting life forms. It has been established by many scientists that freshwater systems like lakes, streams, springs, and other shallow, low-current water bodies comprise about 8% of the uninhabited water regions (dry and terrestrial lands) of the world, house about 6% of all known aquatic organisms.

However, based on the recent trend of emergent pollutants effects on freshwater systems, most of these aquatic plankton resources are threatened and their services to this environment are altered which may pose serious ecological fears for present and future generations.

The book a collection of 12 chapters, from experts around the globe, overviews the impact of emergent pollutants in freshwater plankton communities putting into cognizance the ecological effects and possible sustainable mitigation strategies and management. It will help the academia, researchers, health professionals, social scientists, industrial bodies, and municipalities to spot out the economic, health, and environmental issues caused by the emergent pollutants as a potential threat to life forms.

This book also provides broad glimpses for freshwater ecologists in the in-depth aspects of emerging pollutant compounds and their challenges, the transgenerational and ecological effect of pharmaceuticals and cosmetic wastes on the plankton community, the analytical study of the environmental effects of emergent pollutants on the plankton community, the role of emergent pollutants on the macro-community of planktons (ecosystem influence), the occurrence and fate of pharmaceutical and cosmetic wastes on plankton consortia, the influence of cosmetic and pharmaceutical wastes on the community of plankton, zooplankton and phytoplankton community, influence of personal care products, legislative status, possible remediation, bacteria and algae consortia in aquatic ecosystem and the effects of personal care products on their ecological role, the effects of personal care products on surface water floating organisms, the physical and biological removal of the mass load of emergent pollutants from waste treatment facilities, the economic management and challenges faced in the management of emergent pollutant, and the legislative framework in addressing emergent pollutants and ecological impacts.

About the Book

This book introduces the environmental and health monitoring of emergent pollutants and their influences on the community structure of lentic freshwater planktons.

It highlights some challenges in the improper disposal of industrial pharmaceutical wastes not properly treated and the possible environmental and health risks.

It suggests possible sustainable mitigation techniques in the treatment of the waste pollutants.

The book addresses the issues of the regulatory and monitoring framework, as well as reviews some laws governing the disposal of wastes management.

About the Editors

Osikemekha Anthony Anani is Senior Lecturer in the Department of Animal and Environmental Biology, Delta State University, Abraka, Delta State, Nigeria. His academic qualifications are B.Sc. (Ed.) (Hons.) Biology and Education (2006) in the prestigious Delta State University, Abraka, M.Sc. Environmental Quality Management (2013), and Ph.D. Environmental Quality Management (2017) in the respected University of Benin, Benin City, both in Nigeria.

His areas of proficiency are pollution studies, ecotoxicology, ecology, forensic and aquatic biology, hydrobiology, benthology, malacology, ecological and health risk valuation, and conservation of plants and animals.

Dr. Anani has over 100 peer-reviewed scientific publications in international and national scientific journals, 11 presentations at local and international conferences, 58 Scopus papers, and over 11 Web of Science. His published academic pieces can be found via the following links:

https://scholar.google.com/citations?user=5LvCxrYAAAAJ&hl=en
https://orcid.org/0000-0001-6277-3314
https://www.researchgate.net/profile/Anani_Anthony
https://www.scopus.com/authid/detail.uri?authorId=56470817000
https://loop.frontiersin.org/people/1286170/overview
https://www.webofscience.com/wos/author/search

He leads a research group in the laboratory for ecotoxicology and forensic biology at the same university, which focuses on environmental sustainability and pollution monitoring.

Maulin P. Shah, currently working as Deputy General Manager at the Industrial Wastewater Research Lab, in the Division of Applied and Environmental Microbiology Lab at Enviro Technology Ltd., Ankleshwar, Gujarat, India, received his Ph.D. (2002–2005) in Environmental Microbiology from Sardar Patel University, Vallabh Vidyanagar, Gujarat. He served as Assistant Professor at Gujarat University, Godhra, in 2001. He is a microbial biotechnologist with diverse research interests. A group of research scholars is working under his guidance on the areas ranging from applied microbiology, environmental biotechnology, bioremediation, and industrial liquid waste management to solid state fermentation. His primary interest is the environment, the quality of our living resources, and the ways that bacteria can help to manage and degrade toxic waste and restore environmental health. Consequently, he is very interested in genetic adaptation processes in bacteria, the mechanisms by which they deal with toxic substances, how they react to pollution in general, and how we can apply microbial processes in a useful way (like bacterial bioreporters). One of our major interests is to study how bacteria evolve and adapt to use organic pollutants as novel growth substrates. Bacteria with new degradation capabilities are often selected in polluted environments and have accumulated small (mutations) and large genetic changes (transpositions, recombination, and horizontally transferred elements). His work has been focused on assessing the impact of industrial pollution on microbial diversity of wastewater following cultivation-dependent and cultivation-independent analysis. His major work involves isolation, screening, identification, and genetic engineering of high impact of microbes for the degradation of hazardous materials. He has more than 200 research publications in highly reputed national and international journals. He directs the research program at Enviro Technology Ltd., Ankleshwar. He has guided more than 100 postgraduate students in various disciplines of life science. He is an active editorial board member in more than 150 highly reputed journals in the field of environmental and biological sciences. He was Founder Editor-in-Chief of the *International Journal of Environmental Bioremediation and Biodegradation* (2012–2014) as well as the *Journal of Applied and Environmental Microbiology* (2012–2014) (Science and Education Publishing, USA). He is actively engaged as Editorial Board Member in 32 journals of high repute (*Elsevier, Springer, Taylor & Francis, RSC, Wiley, KeAi, De Gruyter*). He is also serving as a reviewer in various journals of national and international repute. He has edited more than 200 books in wastewater microbiology, environmental microbiology, bioremediation, and hazardous waste treatment.

Contributors

K.K. Adama
Department of Chemical Engineering
Faculty of Engineering
Edo State University Uzairue
Edo State, Nigeria

P.A. Aidonojie
Kampala International University
Kansanga, Kampala, Uganda

I.E. Ainyanbhor
Department of Biochemistry
Faculty of Science
Delta State University of Science and Technology
Ozoro, Delta State, Nigeria

A.G. Anani
Department of Animal and Environmental Biology
Faculty of Life Science
University of Benin
Benin City, Edo State, Nigeria

O.A. Anani
Department of Animal and Environmental Biology
Faculty of Science
Delta State University
Abraka, Delta State, Nigeria

O. Aregbor
Department of Marine Science
Delta State University of Science and Technology
Ozoro, Delta State, Nigeria

O. Aruoren
Department of Biochemistry
Faculty of Science
Delta State University of Science and Technology
Ozoro, Delta State, Nigeria
and
Department of Biological Sciences Faculty of Natural Sciences and Environmental Studies
Godfrey Okoye University
Enugu, Enugu State, Nigeria

O. Ekayoda
Department of Biochemistry
Faculty of Science
Delta State University
Abraka, Delta State, Nigeria

P.M. Etaware
Department of Plant Science and Biotechnology
Faculty of Science
Delta State University of Science and Technology
Ozoro, Delta State, Nigeria

U.F. Evuen
Department of Biochemistry
Faculty of Science
Delta State University of Science and Technology
Ozoro, Delta State, Nigeria

E.M. Eze
Department of Microbiology
Faculty of Science Delta State University of Science and Technology
Ozoro, Delta State, Nigeria

I.H. Innocent
Department of Microbiology
Faculty of Sciences
University of Portharcourt
Rivers State, Nigeria

F.E. Isoje
Department of Science Laboratory Technology, Faculty of Science
Delta State University of Science and Technology
Ozoro, Delta State, Nigeria

O.E. Jessa
Department of Animal and Environmental Biology
Faculty of Science
Delta State University of Science and Technology
Ozoro, Delta State, Nigeria

S. Jothi Murugan
Department of Biotechnology
Rajalakshmi Engineering College
Thandalam, Chennai, Tamil Nadu

L.U. Obi
Microbiology and Environmental Biotechnology Research Group
Agricultural Research Council—Natural Resources and Engineering
Pretoria, South Africa

P.O. Obiebi
Department of Microbiology
Faculty of Science
Delta State University of Science and Technology
Ozoro, Delta State, Nigeria

P.I. Ogwezzy
Department of Animal and Environmental Biology,
Faculty of Science
Delta State University of Science and Technology
Ozoro, Delta State, Nigeria

B.U. Okobia
Department of Microbiology
Faculty of Science
Delta State University of Science and Technology
Ozoro, Delta State, Nigeria

S.U. Okom
Department of Biochemistry
Faculty of Science
Delta State University of Science and Technology
Ozoro, Delta State, Nigeria

J. Okpoghono
Department of Biochemistry
Faculty of Science
Delta State University of Science and Technology
Ozoro, Delta State, Nigeria

F.N. Olisaka
Department of Biological Sciences, Faculty of Natural Sciences and Environmental Studies
Godfrey Okoye University
Enugu, Enugu State, Nigeria

P.E. Omoruwou
Department of Animal and Environmental Biology
Faculty of Science
Delta State University of Science and Technology
Ozoro, Delta State, Nigeria

P. Onyemaechi
Department of Biological Sciences
Faculty of Natural Sciences and Environmental Studies
Godfrey Okoye University
Enugu, Enugu State, Nigeria

F.C. Onyia
Department of Microbiology
Federal University
Oye Ekiti, Ekiti State, Nigeria

J.O. Orogu
Department of Microbiology
Faculty of Science
Delta State University of Science and Technology
Ozoro, Delta State, Nigeria

S.S. Sai Sandhya
Department of Biotechnology
Rajalakshmi Engineering College
Thandalam, Chennai, Tamil Nadu

G. Sarada
Department of Biotechnology
Rajalakshmi Engineering College
Thandalam, Chennai, Tamil Nadu

K.S. Shreenidhi
Department of Biotechnology
Rajalakshmi Engineering College
Thandalam, Chennai, Tamil Nadu

O. Ukolobi
Department of Microbiology
Faculty of Science
Delta State University of Science and Technology
Ozoro, Delta State, Nigeria

B. Vijaya Geetha
Department of Biotechnology
Rajalakshmi Engineering College
Thandalam, Chennai, Tamil Nadu

Acknowledgments

To begin with, we invoke the name of the supreme God—the beneficent and the most merciful—for enabling us to complete this task successfully. We want to thank all the authors that have contributed to this book for their dedicated efforts and excellent contributions.

We also wish to acknowledge the anonymous reviewers for their valuable inputs to enhance the quality of the book in the present form. We are highly thankful to the dedicated editorial team of the Taylor & Francis Group, LLC, State of Delaware, especially Joe Clements and Maggie Apostolis, for their kind support to complete this book project successfully.

We hope that readers will enjoy the book and get enlightened with recent developments related to emergent pollutants in freshwater plankton communities, their ecological effects, management, and sustainable mitigation strategies

Each editor is thankful to their parents, better halves, and kids for their kind support and encouragements.

1 Emerging Pollutant Compounds Challenges for Freshwater Ecologists

Paul Ikechuku Ogwezzy, Peter Mudiaga Etaware, Precious Onome Obiebi, Uduenevwo Francis Evuen, Solomon Ugochukwu Okom, Ebere Mary Eze, Irene Ebosereme Ainyanbhor, Oke Aruoren, Osikemekha Anthony Anani, and Joshua Othuke Orogu

1.1 INTRODUCTION

Emerging pollutants (EPs) manifest in diverse concentrations, displaying seasonal fluctuations across water bodies, underground waters, soils and sediments, biological systems, and ambient air. In the European aquatic environment, more than 700 emerging pollutants have been categorized into twenty (20) classes. However, there is a significant gap in our understanding of their fate, behaviors, and effects, as well as the development of effective treatment technologies for their efficient removal (Gogoi et al., 2018). Some of the emerging pollutants that are predominantly organic compounds include pharmaceutical and personal care products (PPCPs), steroids and hormones, food additives, pesticides, plasticizers, wood preservatives, laundry detergents, disinfectants, surfactants, flame retardants, and other organic compounds in water. These pollutants are primarily generated as a result of human activities (Yankui et al., 2019). For example, the introduction of emerging pollutants into freshwater ecosystems occurs as a result of human activities such as industrial effluent, distribution, utilization, and disposal (Cao et al., 2019; Hashemi and Kaykhaii, 2022). Geissen et al. (2015) revealed that the presence of emerging pollutants (EPs) in the environment can arise from diverse points or diffuse sources. Subsequently, these EPs can infiltrate the soil, atmosphere, or water bodies through multiple pathways or mechanisms, which are primarily influenced by the properties of EPs such as polarity, volatility, persistence, and other relevant factors, as well as the characteristics of the environmental compartments. Since emerging pollutants have negative effects on humans and aquatic life, their presence in freshwater ecosystems is a matter of public concern (Tavengwa and Dalu, 2022; Tavengwa et al., 2022). According to Enyoh et al. (2020),

DOI: 10.1201/9781003362975-1

the emergence of these pollutants poses a direct threat to human health. The potential risks include unconsciousness, headaches, vomiting, skin irritations, coronary heart diseases, damage to the central nervous system and liver, as well as DNA mutation and cancer. It was reported that about 8 million tons of bio-plastic waste were dumped into the ocean, each year, making bio-plastics the most prevalent emerging water pollutant (Kehinde et al., 2019). Giant, large, medium, micro, and nano bio-plastics are classified according to their respective sizes (i.e. giant > 1 m, large < 1 m, medium < 2.5 cm, micro < 5 mm, and nano < 0.1 μm bio-plastic size range, respectively). The scientific understanding of potential health risks to both humans and the environment from emerging freshwater pollutants remains limited, especially in the developing world. Additionally, there is a lack of knowledge regarding the presence of these pollutants in water resources and wastewater, as well as their pathways and accumulation in the environment (Sanganyando and Kajau, 2022). Therefore, it is imperative to advance scientific understanding, especially in the developing world where research is far behind. Appropriate technical and policy approaches should also be adopted to monitor emerging pollutants in wastewater and water resources to avoid or minimize the potential risks to human health and the environment. Also, preventive measures and/or control strategies for EP disposal into water bodies and the environment should be defined, developed, and implemented urgently.

1.2 OCCURRENCE OF EMERGING POLLUTANTS IN WATER BODIES

The presence of emerging pollutants (EPs) in surface water and the aquatic ecosystem is a significant issue, as these water bodies are typically the primary source of drinking water in most rural communities around the world and also the feed-in source (water inlet) for the majority of the public drinking water treatment plants (PDWTPs), thereby posing a threat to public health (Haddaoui and Metao-Sagasta, 2021). In the Middle East and North African (MENA) countries, there have been 57 reported cases of EPs in various fresh and drinking waters, surpassing the thresholds set by the World Health Organization (WHO) and the European Commission (EC) (Haddaoui and Metao-Sagasta, 2021). In marine bivalves collected near point source municipal wastewater and landfill leachate effluent discharges in Hong Kong, low levels of various classes of pharmaceuticals, pesticides, and phosphorus-based flame retardants were detected in μg/kg (Burket et al., 2018). Similar studies conducted in Northwestern Italy showed that drinking water treatment plants were found to contain low concentrations of EPs (Magi et al., 2018).

1.3 EFFECTS OF EMERGING POLLUTANTS ON SURFACE WATER ORGANISMS

Studies have shown that the introduction of new pollutants into aquatic ecosystems can alter both water quality and the composition of the plankton communities (see Table 1.1). Plankton is susceptible to a wide range of impacts from developing contaminants, depending on several variables including the kind of pollutant, its concentration, the length of exposure, and the species' sensitivity. To effectively manage and conserve aquatic ecosystems, we must understand these consequences to evaluate the ecological hazards associated with developing contaminants. In a study by Chia et al. (2021), it was reported that drug exposure had diverse effects on phytoplankton, such as growth inhibition, smaller cell sizes, cell deformities, reduced photosynthesis, and decreased biomass. Jönander et al. (2023) revealed that the discharged water from ships contains metals and organic compounds, including polycyclic aromatic hydrocarbons, which are known for their detrimental effects on marine zooplankton. Exposure to harmful environmental chemicals reportedly causes aquatic organisms to exhibit various immune-related problems, such as shifts in lymphoid tissue organization, changes in immune functionality and cell distribution, and reduced capacity to resist infections. The alteration of immune systems by pollutants could lead to increased mortality among individual organisms, greater vulnerability to pathogen attacks, and potentially diminished populations of specific species (Kataoka and Kashiwada, 2022).

TABLE 1.1
The Following Table Shows the List of Emergent Wastes in Water Bodies, Their Impacts on Water Resources, and Their Mode of Transfer.

Emergent wastes	Impact on water resources	Mode of transfer
Steroids and Hormones	• Hormonal disruption in aquatic organisms	• Discharge from industrial processes
Pharmaceutical Residues	Potential harm to aquatic ecosystems and human health	• Discharged from wastewater treatment plants
	• Antibiotic resistance development in aquatic organisms	• Runoff from pharmaceutical manufacturing facilities
	• Disruption of endocrine systems in aquatic organisms	• Disposal of unused medications
Personal Care Products	• Accumulation in sediments and aquatic organisms	• Direct discharge from households and industries
	• Hormonal disruption in aquatic organisms	• Urban runoff and stormwater
	• Impact on aquatic plant growth and reproduction	
Chemicals (EDCs)	• Altered reproductive behaviors in fish and amphibians	• Agricultural runoff containing pesticides and herbicides
	• Potential impacts on human reproductive health	• Leaching from landfills and waste disposal sites
Microplastics	• Physical harm to aquatic organisms and ecosystems	• Primary sources: microbeads, microfibers, and fragments
	• Bioaccumulation in aquatic food webs	• Secondary sources: breakdown of larger plastic items

Source: Organisation for Economic Co-Operation and Development (2012). *New and Emerging Water Pollutants Arising from Agriculture*. Available at www.oecd.org/agriculture/water.

Furthermore, artificial sweeteners were extensively found in elevated concentrations, within the groundwater and surface water of the Dongxiang River basin in southern China, as reported by Yang et al. (2018). In analyses of rainwater samples collected in Moscow (Russia), researchers detected more than 700 organic compounds from different chemical classes, notably, pyridines, organophosphates, and dichloronitromethane, etc., among other compounds that were classified as EPs (Polyakova et al., 2018). Emerging pollutants are prevalent in freshwater, sewer overflows, and wastewater across various regions, including Europe, North America, Brazil, Australia, India, and China (Hughes et al., 2013; Petrie et al., 2015; Copetti et al., 2019).

1.4 CHALLENGES FACED BY ECOLOGISTS IN REGULATING EMERGING CONTAMINANTS IN FRESHWATER SYSTEMS

The regulation of emerging contaminants in freshwater systems poses several challenges, reflecting the dynamic nature of these contaminants and the evolving understanding of their potential impacts on human health and the environment. Ecologists face several challenges in regulating emerging pollutants in freshwater systems. These challenges arise from the complexity of emerging pollutants, their dynamic interactions with the environment, and the limitations of control and regulatory frameworks. Some of the key challenges include the following.

1.4.1 Dispersion, Transformation, and Accumulation of Emerging Pollutants

In rural regions, emerging pollutants (EPs) are widely dispersed throughout the area and are transported through various means such as air, runoff, erosion, or leaching until they reach a water body (Ter-Laak et al., 2006). The adsorption behavior of pharmaceuticals, for instance, can significantly differ in different soil types due to their occurrence in both ionized and un-ionized forms. This variation affects their interaction with different compounds present in the soil (Ter Laak et al., 2006). The living environment may be significantly influenced by the properties of intermediates and/or end products resulting from the photodegradation of EPs, as stated by Xie et al. (2014). EPs, which include chemicals and microbial agents, are often nondegradable and persistent. As a result, they can travel through various environmental matrices via water and air, as noted by Rasheed et al. (2021). Furthermore, these contaminants can be found in areas where they were previously absent. These pollutants can bioaccumulate and biotransform in aquatic organisms, causing the formation of metabolites or deposition of these substances in tissues. Some of these aquatic organisms include salmon, mussels, oysters, shrimp, crabs, frogs, salamanders, seabirds, and fungi.

1.4.2 Insufficient Technical Ability to Evaluate, Track, and Register Synthetic Materials

Ecologists and other regulatory agencies around the world, including the Pesticide Control Board in developing countries, face challenges in monitoring and maintaining synthetic substance inventories due to inadequate technical infrastructure

(Kosamu et al., 2020) (see Table 1.2). This often results in the inability to conduct pesticide residue analysis in the environment and control illegal importation. According to a survey conducted by the World Health Organization in 2014, half of the African nations lacked the necessary laboratory infrastructure to evaluate, supervise, or identify emerging pollutants in environmental matrices (WHO Regional Office for Africa, 2014). Many of these developing countries lacked essential facilities for assessing the emerging pollutant. It is crucial to enhance the capability of performing chemical and biological analyses to enhance emissions monitoring and product registry in developing nations. According to the European Environment Agency assessment in 2018, 14 member states reported that between 75% and 99% of the total area of groundwater bodies were in good quantitative status, while three gave a figure of between 50% and 75%.

1.4.3 Poor Science and Policy Collaboration

The participation of scientists in policymaking has been hindered by various factors, as identified by the International Panel on Chemical Pollution. These factors include intrinsic factors such as varying goals on science, different approaches to pertinent questions, diverse perceptions on time frames and languages, and extrinsic factors such as inadequate training, limited access to relevant information, absence of communication platforms, and differing value perceptions (IPCP, 2018). A need for addressing these factors is crucial to ensure effective collaboration between scientists and policymakers. The lobbying efforts of chemical manufacturers frequently worsen the distrust of scientific evidence. A recent study discovered that industry lobbying, along with the absence of local scientific evidence and limited public awareness played a significant role in the absence of national regulations on nanoparticles (Anderson et al., 2012). Similarly, in the United States, an analysis of the failure of a climate policy aimed at regulating greenhouse gases revealed that political lobbying from industries anticipating

TABLE 1.2
The Following Table Shows the Proportion of Good Quantitative Status of Groundwater Bodies by Area.

Member States	% of groundwater bodies in good quantitative status, by area
Austria, Latvia, Luxembourg, Netherlands, Romania, Slovenia	100%
Croatia, Denmark, Estonia, Bulgaria, Portugal, Germany, Poland, Finland, Sweden, Czech Republic, France, United Kingdom, Spain, Italy	75%–100%
Hungary, Slovakia, Belgium	50%–75%
Cyprus, Malta	< 50%

Source: WISE-SoW database data from 25 member states (EU-28 except for Greece, Ireland, and Lithuania).

losses greatly contributed to the policy's downfall (Meng and Rode, 2019). The lack of sufficient science-policy interface forums could be considered the underlying cause for the imbalance observed in the lobbying process. Following the abuse of water bodies, several rules and regulations safeguarding water bodies have been established. United States Environmental Protection Agency (EPA) mandated in cooperation with federal state and local agencies and industries to develop programs for preventing, reducing, or eliminating the pollution of the navigable waters and groundwaters and improving the sanitary condition of surface and underground waters. The Clean Water Act in the United States regulates the discharge of pollutants into US waters. A framework for the protection of groundwater, coastal waters, transitional waters, and inland surface waters was established by the EU Water Framework Directive. The EU directive seeks to safeguard and enhance the aquatic environment within the member states. Generally, there is a high level of adherence among regulated entities due to the legal consequences of non-compliance.

1.5 INADEQUATE INFRASTRUCTURE FOR REGULATING SYNTHETIC SUBSTANCES

Many nations have distinct national organizations responsible for assessing applications for the registration and authorization of various synthetic substances (Wang et al., 2020). Often, there are distinct catalogs for different categories, such as cosmetics, pesticides, pharmaceuticals, food additives, and industrial chemicals, which may be overseen by different regulatory bodies. Many countries and regional bodies have implemented regulatory frameworks to evaluate and oversee the safety of synthetic substances throughout their entire lifespan. These frameworks include a comprehensive record of the chemicals produced, imported, or utilized within the nation. Thousands of synthetic substances inadvertently enter the aquatic environment during use and disposal, as a result of illegal activities, manufacturer confidentiality practices, and inadequate regulatory infrastructure (Kosamu et al., 2020). These substances, often referred to as "ghost" synthetic substances, manage to pass through regulatory cracks. These synthetic substances can enter the aquatic environment through point source and non-point source (see Figure 1.1). While point source pollution has been effectively controlled in recent years, non-point source pollution has been reported to be one of the most important sources of threats to the safety of water resources and has even contributed to the outbreak of water blooms.

The ecological role of emerging pollutants and their mechanisms of action in aquatic biotas can vary depending on the type of pollutants and the specific ecological context. Here are general insights into the ecological roles and mechanisms of action of emerging pollutants in aquatic ecosystems.

1.5.1 Bioaccumulation and Biomagnification

Role: Emerging pollutants, such as pharmaceuticals and certain chemicals, can accumulate in aquatic organisms.

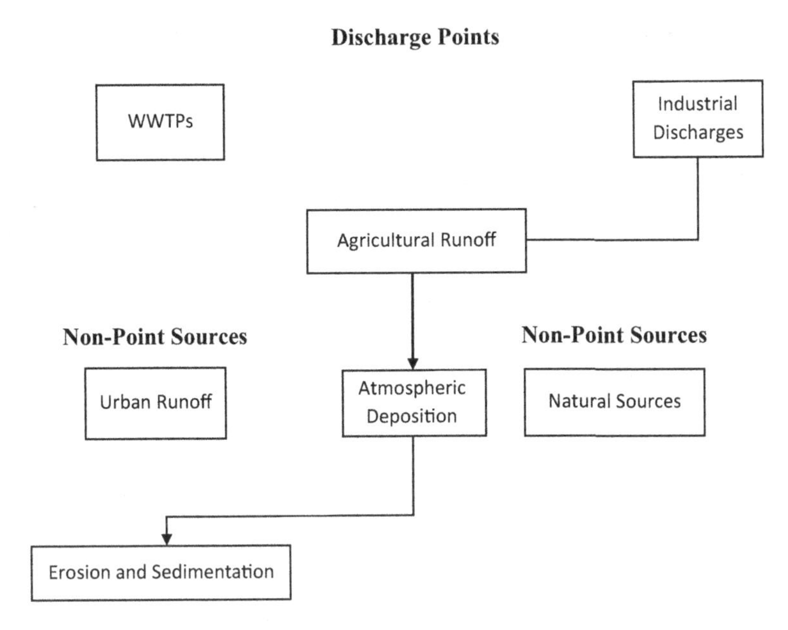

FIGURE 1.1 Review of risk from potential emerging contaminants in UK groundwater.

Source: Stuart et al. (2012).

Mechanism: Bioaccumulation occurs as pollutants are absorbed by aquatic organisms from water and sediments. Biomagnification then occurs as pollutants move up the food chain, reaching higher trophic levels (Kumar et al., 2022).

1.5.2 Endocrine Disruption

Role: Some emerging pollutants, like endocrine-disrupting chemicals (EDCs), can interfere with the endocrine systems of aquatic organisms.

Mechanism: EDCs mimic or interfere with natural hormones, leading to disruptions in reproductive and developmental processes in aquatic biotas (Silva et al., 2018).

1.5.3 Genotoxic Effects

Role: Certain emerging pollutants, including some pharmaceuticals and industrial chemicals, may exhibit genotoxic effects in aquatic organisms.

Mechanism: These pollutants can damage the genetic material of organisms, potentially leading to mutations and other adverse effects on population dynamics (Islam et al., 2022).

1.5.4 Alteration of Behavior and Physiology

Role: Emerging pollutants can influence the behavior and physiological functions of aquatic organisms.

Mechanism: Substances like pharmaceuticals and personal care products may affect the nervous system, altering behaviors such as feeding, mating, and predator avoidance (Weis et al., 2021).

1.5.5 Microbial Community Dynamics

Role: Emerging pollutants can impact the structure and function of microbial communities in aquatic environments.

Mechanism: Microorganisms may be directly affected by pollutants, leading to shifts in community composition and potentially altering nutrient cycling and other ecosystem processes (Gomes et al., 2020).

1.5.6 Antibiotic Resistance Development

Role: The release of antibiotics into aquatic systems as emerging pollutants can contribute to the development of antibiotic resistance in bacteria.

Mechanism: Exposure to low levels of antibiotics may promote the selection of antibiotic-resistant strains of bacteria in aquatic environments (Niu et al., 2022).

1.5.7 Disruption of Trophic Interactions

Role: Emerging pollutants can disrupt trophic interactions in aquatic food webs.

Mechanism: Changes in the abundance or behavior of certain species due to pollutant exposure can have cascading effects on predator-prey relationships and overall ecosystem structure (Samantha et al., 2019).

1.5.8 Impact on Reproductive Success

Role: Emerging pollutants, especially those with endocrine-disrupting properties, can affect reproductive success in aquatic organisms.

Mechanism: Adverse effects on reproductive organs and processes may lead to reduced reproductive success, potentially influencing population dynamics (Ma et al., 2019).

1.5.9 Changes in Community Composition

Role: Emerging pollutants can contribute to shifts in the composition of aquatic communities.

Mechanism: Some organisms may be more tolerant to pollutants, leading to changes in the abundance and distribution of species within aquatic ecosystems (Xin et al., 2020).

The presence of emerging pollutants (EPs) in surface water and the aquatic ecosystem is a major cause of global concern. Several limitations affecting the regulation and control of the disposal of EPs were highlighted in this chapter. Also, the challenges faced by the global regulatory bodies and ecologists around the world were itemized, analyzed, and discussed herein. The fact remains that EPs are fast becoming a formidable threat to all forms of life and the environment as a whole.

1.6 CONCLUSION

There is an urgent need for the global community to define, develop, and implement efficient methods for the management of emerging pollutants from indiscriminately disposed wastes all around the world. Also, severe measures should be put in place to curtail lobbying and other crimes committed by product owners and manufacturers. In Nigeria, indiscriminate disposal of unused and expired medications, which can become potentially toxic substances when they accumulate in various matrices such as water and soil, should be controlled using modern techniques. Furthermore, there exists a dearth of knowledge regarding the techniques and procedures employed by community pharmacists in Nigeria for disposing of expired medications culminating in the increase in emergent pollutants in the environment.

REFERENCES

Anderson, P.D., Denslow, N.D., Drewes, J.E., Olivieri, A.W., Schlenk, D., Scott, G.I., and Snyder, S. (2012). *Monitoring strategies for chemicals of emerging concern (CECs) in California's aquatic ecosystems: Recommendations of a science advisory panel*. Technical Report 692. Southern California Coastal Water Research Project, Costa Mesa, CA.

Burket, S., Sapozhnikova, Y.V., Zheng, J., Chung, S. and Brooks, B.W. (2018). At the intersection of urbanization, water, and food security: Bioaccumulation of select contaminants of emerging concern in mussels and oysters from Hong Kong. *Journal of Agricultural and Food Chemistry*, **66**, 5009–5017.

Cao, J., Sanganyado, E., Liu, W., Zhang, W. and Liu, Y. (2019). Decolorization and detoxification of direct blue 2B by indigenous bacterial consortium. *Journal of Environmental Management*, **242**, 229–237.

Chia, M., Lorenzi, A.S., Ameh, I. and Dauda, S. (2021). Susceptibility of phytoplankton to the increasing presence of active pharmaceutical ingredients (APIs) in the aquatic environment: A review. *Aquatic Toxicology*, **234**(7), 105809.

Copetti, D., Marziali, L., Viviano, G., Valsecchi, L., Guzzella, L., Capodaglio, A.G., Tartari, G., Polesello, S., Valsecchi, S., Mezzanotte, V. and Salerno, F. (2019). Intensive monitoring of conventional and surrogate quality parameters in a highly urbanized river affected by multiple combined sewer overflows. *Water Supply*, **19**(3), 953–966.

Enyoh, C.E., Shafea, L., Verla, A.W., Verla, E.N., Wang, Q., Tanzin, C. and Marcel, P. (2020). Microplastics exposure routes and toxicity studies to ecosystems: An overview. *Environmental Analysis Health and Toxicology*, **35**(1), 1–10.

European Environment Agency. (2018). *European Waters Assessment of Status and Pressures 2018*. Available at https://www.eea.europa.eu/publications/state-of-water. (Accessed: 30th December 2023).

Geissen, V., Mol, H., Klumpp, E., Umlauf, G., Nadal, M., Van der Ploeg, M., Van de Zee, S.E.A.T.M. and Ritsemaa, C.J. (2015). Emerging pollutants in the environment: A challenge for water resource management. *International Soil and Water Conservation Research*, **3**, 57–65.

Gogoi, A., Mazumder, P., Tyagi, V.K., Tushara Chaminda, G.G., An, A.K., and Kumar, M. (2018). Occurrence and fate of emerging contaminants in water environment: A review. *Groundwater for Sustainable Development*, **6**, 169–180. doi:10.1016/j.gsd.2017.12.009

Gomes, I.B., Maillard, J., Simões, L.C. and Simões, M. (2020). Emerging contaminants affect the microbiome of water systems—strategies for their mitigation. *Nature Partner Journals Clean Water*, **3**(39).

Haddaoui, I. and Metao-Sagasta, J. (2021). A review on occurrence of emerging pollutants in waters of the MENA region. *Environmental Science and Pollution Research International*, **28**(48), 68090–68110.

Hashemi, S.H. and Kaykhaii, M. (2022). Azo dyes: Sources, occurrence, toxicity, sampling, analysis and their removal methods. In: Dalu, T. and Tavengwa, N.T. (Eds.), *Emerging Freshwater Pollutants: Analysis, Fate and Regulations*. Elsevier, Cambridge.

Hughes, S.R., Kay, P. and Brown, L.E. (2013). Global synthesis and critical evaluation of pharmaceutical data sets collected from river systems. *Environmental Science and Technology*, **47**, 661–677.

IPCP. (2018). *Strengthening the Science-Policy Interface in International Chemicals Governance: A Thought Starter on Options for a Way Forward*. Available at https://www.ipcp.ch/wp-content/uploads/2019/02/IPCP-Sci-Pol-WorkshopThoughtStarter2018.pdf (Accessed: 5th October 2023).

Islam, A., Amin, S.N.M., Rahman, M.A., Juraimi, A.S., Uddin, K., Brown, C.L. and Arshad, A. (2022). Chronic effects of organic pesticides on the aquatic environment and human health: A review. *Environmental Nanotechnology, Monitoring and Management*, **18**(22), 100740.

Jönander, C., Egardt, J., Hassellöv, I.-M., Tiselius, P., Rasmussen, M. and Dahllöf, I. (2023). Exposure to closed-loop scrubber washwater alters biodiversity, reproduction, and grazing of marine zooplankton. *Frontiers in Marine Science*, **10**, 1249964.

Kataoka, C. and Kashiwada, S. (2022). Ecological risks due to immunotoxicological effects on aquatic organisms. *International Journal of Molecular Sciences*, **22**(15), 8305.

Kehinde, O., Omotosho, O.A. and Ohijeagbon, I.O. (2019). The Effect of Varying Sand and Plastic Additives on The Mechanical Properties of Cement Matrix Tiles. *Journal of Physics: Conference Series*. 1378: 2. https://iopscience.iop.org/article/10.1088/1742-6596/1378/2/022077.

Kosamu, I., Kaonga, C. and Utembe, W. (2020). A critical review of the status of pesticide exposure management in Malawi. *International Journal of Environmental Research and Public Health*, **17**, 1–13.

Kumar, J.A., Krithiga, T., Sathish, S.A., Renita, A., Prabu, D., Lokesh, S., Geetha, R., Karthic, S., Namasivayam, R. and Sillanpaa, M. (2022). Persistent organic pollutants in water resources: Fate, occurrence, characterization and risk analysis. *Science of the Total Environment*, **831**(22), 154808.

Ma, Y., He, X., Qi, K., Wang, T., Qi, Y., Cui, L., Wang, F. and Song, M. (2019). Effects of environmental contaminants on fertility and reproductive health. *Journal of Environmental Sciences*, **77**(19), 210–217.

Magi, E., DiCarro, M., Mirasole, C. and Benedetti, B. (2018). Combining passive sampling and tandem mass spectrometry for the determination of pharmaceuticals and other emerging pollutants in drinking water. *Microchemical Journal*, **136**, 56–60.

Meng, K.C. and Rode, A. (2019). The social cost of lobbying over climate policy. *Nature Climate Change*, **9**(6), 472–476.

Niu, L., Liu, W., Juhasz, A., Chen, J. and Ma, L. (2022). Emerging contaminants antibiotic resistance genes and microplastics in the environment: Introduction to 21 review articles published in CREST during 2018–2022. *Critical Reviews in Environmental Science and Technology*, **52**(22), 4135–4146.

Organisation for Economic Co-Operation and Development. (2012). *New and Emerging Water Pollutants Arising from Agriculture*. Available at www.oecd.org/agriculture/water (Accessed: 26th December 2023).

Petrie, B., Barden, R. and Kasprzyk-Hordern, B. (2015). A review on emerging contaminants in wastewaters and the environment: Current knowledge, understudied areas and recommendations for future monitoring. *Water Research*, **72**, 3–27.

Polyakova, O.V., Artaev, V.B. and Lebedev, A.T. (2018). Priority and emerging pollutants in the Moscow rain. *Science of the Total Environment*, **645**, 1126–1134.

Rasheed, T., Rizwan, K., Bilal, M., Sher, F. and Iqbal, H. (2021). Tailored functional materials as robust candidates to mitigate pesticides in aqueous matrices-a review. *Chemosphere*, **282**, 131056. https://doi.org/10.1016/j.chemosphere.2021.131056

Samantha, C.P., Samantha, M.D., Stephanie, L.R., Krista, A.M. and Michael, W.M. (2019). *Life in a Contaminant Milieu: PPCP Mixtures Generate Unpredictable Outcomes Across Trophic Levels and Life Stagest*. Available at http://hdl.handle.net/10342/7582 (Accessed: 27th December 2023).

Sanganyando, E. and Kajau, T. (2022). The fate of emerging pollutants in aquatic systems: An overview. Chapter 7. In: *Emerging Freshwater Pollutants. Analysis, Fate and Regulations*. Elsevier, pp. 119–135. https://doi.org/10.1016/B978-0-12-822850-0.00002-8.

Silva, A.P.A., Oliveira, C.D.L., Quirino, A.M.S., Silva, F.D.M. and Saraiva, R.A. (2018). Endocrine disruptors in aquatic environment: Effects and consequences on the biodiversity of fish and amphibian species. *Aquatic Science and Technology*, **6**(1), 35.

Stuart, M., Lapworth, D., Crane, E. and Hart, A. (2012). Review of risk from potential emerging contaminants in UK groundwater. *Science of the Total Environment*, **416**, 1–21.

Tavengwa, N.T. and Dalu, T. (2022). Introduction to emerging freshwater pollutants. In: Dalu, T. and Tavengwa, N.T. (Eds.), *Emerging Freshwater Pollutants: Analysis, Fate and Regulations*. Elsevier, Cambridge.

Tavengwa, N.T., Moyo, B., Musarurwa, H. and Dalu, T. (2022). Future directions of emerging pollutants research. In: Dalu, T. and Tavengwa, N.T. (Eds.), *Emerging Freshwater Pollutants: Analysis, Fate and Regulations*. Elsevier, Cambridge.

Ter-Laak, T.L., Agbo, S.O., Barendregt, A. and Hermens, J.L. (2006). Freely dissolved concentrations of PAHs in soil pore water: Measurements via solid-phase extraction and consequences for soil tests. *Environmental Science and Technology*, **40**, 1307–1313.

Wang, M., Cao, R., Zhang, L., Yang, X., Liu, J., Xu, M., Shi, Z., Hu, Z., Zhong, W. and Xiao, G. (2020). Remdesivir and chloroquine effectively inhibit the recently emerged novel coronavirus (2019-nCoV) in vitro. *Cell Research*, **30**, 269–271.

Weis, J.S., Smith, G., Zhou, T., Santiago-Bass, C. and Weis, P. (2021). Effects of contaminants on behavior: Biochemical mechanisms and ecological consequences: Killifish from a contaminated site are slow to capture prey and escape predators; altered neurotransmitters and thyroid may be responsible for this behavior, which may produce population changes in the fish and their major prey, the grass shrimp. *BioScience*, **51**(3), 209–217.

WHO. (2014). *Chemicals of Public Health Concern and Their Management in the African Region, Afro Library Cataloguing-in-Publication Data*. Available at https://www.afro.who.int/sites/default/files/2017-06/9789290232810.pdf (Accessed: 15th October 2023).

Xie, S., Wu, Y., Wang, W., Wang J., Luo, Z. and Li, S. (2014). Effects of acid/alkaline pretreatment and gamma-ray irradiation on extracellular polymeric substances from sewage sludge. *Radiation Physics and Chemistry*, **2014**, 349–353.

Xin, T., Li, R., Zhang, B., Yu, H., Kong, X., Bai, Z., Deng, Y., Jia, L. and Jin, D. (2020). Pollution gradients altered the bacterial community composition and stochastic process of rural polluted ponds. *Microorganisms*, **8**(2), 311.

Yang, Y.Y., Zhao, J.L., Liu, Y.S., Liu, W.R., Zhang, Q.Q., Yao, L. and Ying, G.G. (2018). Pharmaceuticals and personal care products (PPCPs) and artificial sweeteners (ASs) in surface and ground waters and their application as indication of wastewater contamination. *Science of the Total Environment*, **616–617**, 816–823.

Yankui, T., Yin, M., Yang, W., Li, H., Zhong, Y., Mo Liyang, Y., Ma, X. and Sun, X. (2019). Emerging pollutants in water environment: Occurrence, monitoring, fate, and risk assessment. *Water Environment Research*, **91**, 984–991.

2 Transgenerational and Ecological Effect of Pharmaceuticals and Cosmetic Wastes on Plankton Community

Joel Okpoghono, Ovigueroye Ekayoda, Osikemekha Anthony Anani, Fegor Endurance Isoje, Ebere Mary Eze, Ochuko Ernest Jessa, Joshua Othuke Orogu, Solomon Ugochukwu Okom, and Bishop Uche Okobia

2.1 INTRODUCTION

Research indicates that PPCPs negatively affect phytoplankton, which, therefore, negatively affects creatures at higher trophic levels (Gomaa et al., 2020). PPCPs are found in surface water and sediment all over the world and are typically released to the environment by solid waste leachate, industrial wastewater, wastewater from aquaculture, and home wastewater, among other means (Kallenborn et al., 2018). There have been multiple cosmetics and pharmaceutical waste originating in the environment. Pharmaceuticals frequently seen in the environment are shown in Figure 2.1. Prescription and nonprescription pharmaceuticals for humans, illicit substances, and veterinary medications are all included in PPCPs. Additionally, their metabolites and conjugates, for instance, hormones, anticonvulsants, antibiotics, antidepressants, lipid regulators, nonsteroidal anti-inflammatory medicines, and antihypertensives, are included (Roveri et al., 2022). A significant amount of PPCPs have enthralled the sediments in rivers and the majority of planktonic populations as a result of the continual PPCPs introduction in the aquatic environment (Gomaa et al., 2020).

Evolutionary toxicology entails inheritable and transgenerational impacts of pollution exposure that are not always expected from the contaminants' modes of toxicity. Populations vulnerable to chemical pollution have shown genetic reactions, and the effects of transgenerational and evolutionary processes may change

DOI: 10.1201/9781003362975-2

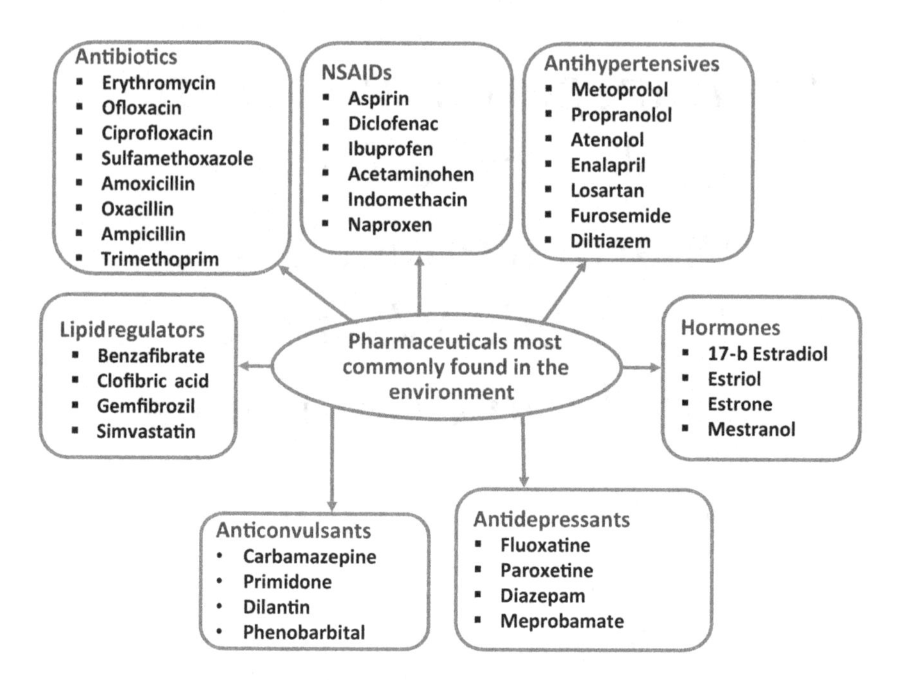

FIGURE 2.1 Pharmaceuticals frequently detected in the environment.

Source: Magureanu et al. (2015).

population dynamics along with the structure and functioning of ecosystems (Bickham, 2011). Some issues are currently associated with ecological risk assessment (ERA), such as the perception of how reactions to contaminants increase in complexity as a function of biological level from individuals to the total ecosystems, the necessity for methods that can make exposure and impact assessments more ecologically realistic, and the inexplicable significance of transgenerational responses and effects in comparison to immediate responses and effects. Ecology and evolution are closely interrelated, as the impact of natural selection and gradient are established by ecological settings, while natural ecological settings and populations are shaped by evolutionary settings (Pomati and Nizzetto, 2013).

The persistent presence of several chemical contaminants at sub-lethal concentrations in the environment is due to human activities which are stressors to the ecosystem, and their interactions with intricate biological entities remain largely unexamined. Exposure can result in cumulative impacts that only show up at larger temporal and biological complexity scales as a result of the interplay between physiological, ecological, and evolutionary processes (Rodea-Palomares et al., 2015). The practices of ecological risk assessment are essential components of the sustainable management of chemicals.

PPCPs impact phytoplankton's growth, photosynthesis, cellular metabolism, and community organization. Zooplankton are affected as they depend on phytoplankton for survival (Vaghela et al., 2023). Ingestion may result in zooplankton digestive tract mechanical injury, feeding problems, malnutrition, negative effects on gene expression, excretion, and a rise in zooplankton mortality (Tang et al., 2019). Phytoplankton's contribution to the biogeochemical cycle of several crucial nutrients, energy transmission, and the ecological balance of aquatic habitats is limited by these physiological impacts (Chia et al., 2021). The planktonic populations encounter several chemicals like PPCPs in the aquatic environment. They are usually affected negatively via anthropogenic environmental changes that occur from the PPCP's impacts on the ecological system. This occurs in diverse forms, and the transgenerational responses and effects act on long temporal scales (Chia et al., 2021). The ecological and transgenerational responses can be widespread in response to anthropogenic environmental change. Therefore, the purpose of this book chapter is to review the ecological and transgenerational impacts of PPCP waste on the plankton population.

2.2 EFFECT OF PHARMACEUTICALS AND COSMETIC WASTES ON PLANKTON COMMUNITY

It has been documented that pharmaceutical residues can alter phytoplankton growth. Ekta and Pammi (2018) found that at a dose range of 0.05 to 0.93 mg/L, streptomycin inhibited the growth of blue-green algae. In a brief analysis, Maria (2017) discovered that the harmful effects of pharmaceutical residues released into the water on marine amphipods, or *Ampithoe valida*, increased dramatically with the length of contact with trash concentrations. *Ampithoe valida* was observed to have worse survival rates when exposed to more than 1% of pharmaceutical residues in the control groups. Waterhouse, damselfly larvae, and ramshorn snails were found in aquatic invertebrates that absorbed pharmaceutical chemicals. *Temora turbinata*, a calanoid copepod, has been shown to cause abnormal growth patterns, reduced adult size, decreased egg development, and decreased adult size when in contact with pharmaceutical waste concentrations of more than 1 ppm (Kayode-Afolayan et al., 2022). Regarding plastic packaging, one of the biggest pollutants worldwide is the cosmetics sector. Significant effects were seen on the functional and taxonomic diversity of the natural planktonic community by both the entire mixture of chemicals leached from polyvinyl chloride (PVC) and zinc (stabilizer used to manufacture a range of paste PVC). The photosynthetic efficiency, diversity, and abundance of prokaryotic and eukaryotic microbial primary producers—which are the building blocks of the marine food web—were all dramatically decreased by exposure to plastic leachates. Additionally, significant amounts of chemical additives from cosmetic wastes seep into the ocean, potentially affecting important populations of marine planktonic bacteria. When exposed to increased amounts of PVC leachate, key heterotrophic species, like SAR11 bacteria, the most prevalent planktonic species in water, also showed notable decreases in relative abundance. However, when in contact with plastic garbage, the relatively large quantity of numerous oligotrophic bacteria—including those belonging to the *Alteromonadales*—rose considerably. Suggesting that exposure to PVC leachates might be advantageous for certain microorganisms.

2.2.1 Phytotoxic Mode of Action of PPCPs in Aquatic Plants

Numerous varieties of plants, including algae and aquatic plants may take up, accumulate, or metabolize PPCPs. Given that cell membranes lack PPCP-specific transporters, plants absorb water and these compounds primarily through passive diffusion through their roots (Wei et al., 2023), which is probably prevailing even though some uptake is influenced by energy-consuming processes. This viewpoint has been substantiated by the Gomes et al. (2017) study. In this investigation, the concentration of ciprofloxacin and plant cell membrane permeability showed a positive correlation. Generally speaking, a variety of intricate biological, chemical, and physical processes and variables influence the uptake of PPCPs. For instance, biological changes in components of the cytoplasmic membranes of different plant species/tissues cause PPCPs to exhibit an order of magnitude variances in affinity and permeability through lipid membranes (Wei et al., 2023). The pathways of PPCP metabolism and potential phytotoxicity in aquatic plants are shown in Figure 2.2. Since the majority of PPCPS are bioactive substances, the bioaccumulation and biotransformation of PPCP compounds in plants (via Phase I and II modification) alters plant physiology and important biochemical processes (Figure 2.2), as well as complicates the plant's exogenous metabolism (Gomaa et al., 2020; Wei et al., 2023). The Phase II metabolites move to the vacuole which then undergo separation/isolation (Phase III). In general, plant development and improvement may be directly impacted by contact with PPCPs, such as by direct plant damage and reduced photosynthesis (Wei et al., 2023).

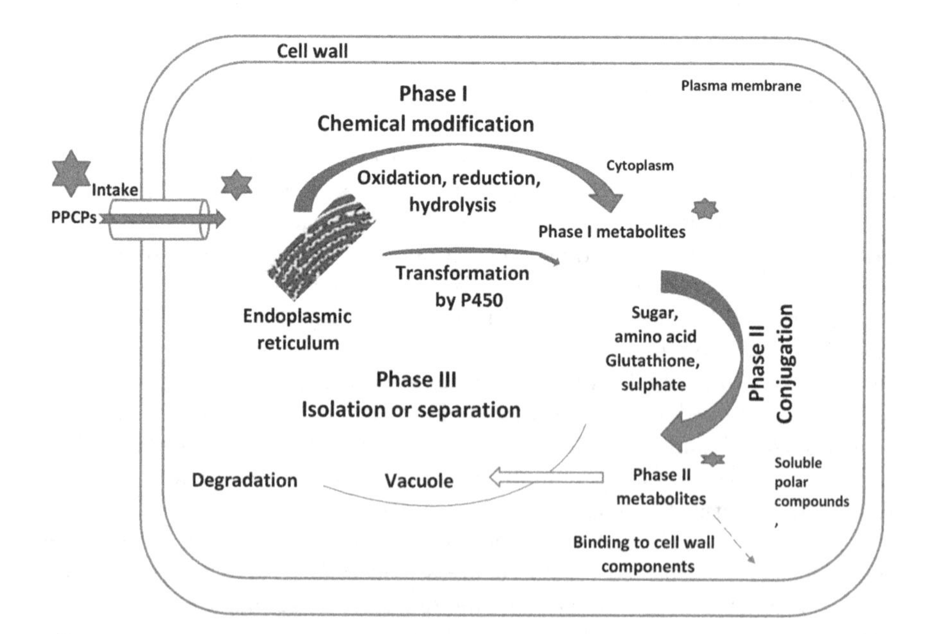

FIGURE 2.2 PPCP metabolism routes and possible phytotoxicity in aquatic plants.

Source: Wei et al. (2023).

2.3 TRANSGENERATIONAL EFFECT OF PPCPS ON PLANKTON COMMUNITY

2.3.1 Altered Community Structure

The effects of PPCPs on the plankton community throughout generations entail the long-term impacts of PPCP exposure that affect not only the populations initially exposed but also subsequent generations within the plankton community. Plankton, which forms the foundation of aquatic food webs, is a crucial part of aquatic ecosystems. They come into contact with PPCPs when released into bodies of water (Gomaa et al., 2020). PPCPs' impacts on plankton can ripple throughout the food chain. Fish and other aquatic animals are among the creatures that might be affected by changes in plankton numbers (Zhou et al., 2020). Plankton can accumulate PPCPs by direct uptake or by consuming contaminated food particles, and the structure and dynamics of entire aquatic communities may change as a result.

Every aquatic environment contains algae (phytoplankton), which presents opportunities for PPCP contact when released. PPCPs tend to accumulate in algae as a result; prolonged exposure of algae to PPCPs causes direct harm to their cell surfaces, including oxidative stress and membrane damage. The life of higher trophic organisms would be impacted by the food web if PPCPs have unanticipated effects on algae. The community mix of phytoplankton and higher trophic species may alter if algal biomass declines, which would likely have an impact on aquaculture and fishing sectors (Xin et al., 2020). PPCP pollutants impact multi-generations of phytoplankton via their cell density, growth rate, yield, cell size, morphology, and ultrastructure. In the interaction between algae and pollutants, morphology and ultrastructure are life-history features (Schaap et al., 2016). Population-level decreases and even isolated extinctions are caused by this interaction.

Plankton population genotoxicity and biomolecular alterations are also brought on by PPCP exposure. Certain PPCPs are genotoxic substances that can change the genetic pool and influence the frequency of mutations. These modifications might impact ecosystem sustainability and population number (Xin et al., 2020). A disruption in normal homeostasis may arise from phytoplankton exposure to genotoxic substances over an extended period, leading to permanent harm and perhaps death (Romero-Murillo et al., 2023).

2.3.2 Transfer Effects

Transgenerational effects describe alterations in an organism's phenotypic or ability to reproduce that are carried on to succeeding generations even in the absence of ongoing exposure to the pollutants. Studies have indicated that plankton may have transgenerational effects from exposure to PPCPs (Zhou et al., 2020). These effects may include changes in neurological function, growth rates, reproductive success, behavior, physiology, and endocrine disruption that are transmitted to future generations. The possible transgenerational toxicity of PPCPs to the planktonic community includes the transfer of the parent's stored PPCPs to the egg and then placed in the progeny. Epigenetic changes, which transform the expression of a gene without changing the underlying DNA sequence, may be the cause of these effects. Variations in the patterns of gene expression point to the possibility of epigenetic alterations (Kumar and Sharma, 2023).

The capacity of aquatic species to reproduce is one of the most prominent transgenerational effects. PPCPs have a direct impact on the development and growth of the offspring by influencing the quantity of sex hormones in the parents or their capacity to reproduce (Zhou et al., 2020). Exposure to PPCPs lowers fertility and raises death rates in exposed plankton communities (Chen et al., 2019). Moreover, PPCPs may influence aquatic creatures' behavior over generations, and it can cause significant physiological and biochemical changes in both parents and offspring which can lead to cross-transgenerational toxicity (Lee et al., 2019). These results were ascribed to variations in the manifestation of the relevant genes. PPCP exposure lowers immunological function in zooplankton progeny, and these effects persisted even after three generations of exposure (Kumar and Sharma, 2023). Certain PPCPs possess the ability to interact with an organism's hormone systems affecting their ability to maintain a healthy population since they are known to be endocrine disruptors (Ebele et al., 2017). Plankton reproductive and developmental problems resulting from endocrine disruption can impact not only the exposed generation but also their progeny (Vera-Chang et al., 2019).

2.4 ECOLOGICAL EFFECT OF PPCPs ON PLANKTON COMMUNITY

2.4.1 Changes in Trophic Interactions

Fish and zooplankton at higher trophic levels may not have access to food if some plankton species are adversely affected by PPCPs. The variety and quantity of plankton species in a community can be impacted by PPCPs. Environmentally relevant amounts of PPCPs may modify ecological connections and processes leading to sublethal ecological consequences of PPCPs which may change the dynamics between different species. For instance, oxazepam, a popular antidepressant, changes how European perch feeds (Brodin et al., 2014). Ecological systems can be permanently disrupted by changing one or more procedures inside the ecosystem. Changes in phytoplankton's growth, photosynthesis, cellular metabolism, community structure, biogeochemical processes, invertebrate development and population dynamics, and other aspects of ecosystem structure and function are facilitated by PPCPs (Wang et al., 2017). PPCPs' contribution to the biogeochemical cycle of several crucial nutrients, energy transmission, and the ecological balance of aquatic habitats is limited by these physiological impacts.

Plankton plays a dynamic role in nutrient cycling in aquatic ecosystems. Changes in their quantity or composition and size-abundance relationships may affect the dynamics of nutrients and alter the ecosystem's health, function, and water quality (Baho et al., 2019). Given the importance of phytoplankton in carbon sequestration and nutrient cycling, changes in community size spectra as a result of PPCP exposure might have far-reaching consequences. PPCP impact can spread throughout trophic levels, and their detrimental effects on the most abundant size class of the phytoplankton community may reduce the effectiveness of the carbon transfer mediated by the phytoplankton food web (Srain et al., 2021). As PPCPs hurt plankton, they may also affect other elements of the ecosystem that rely on plankton for food, such as fish populations, birds, and mammals. The equilibrium of the entire ecosystem may be upset by this.

2.4.2 Bioaccumulation and Biomagnification

The persistence of PPCPs in aquatic habitats results in long-term exposure leading to bioaccumulation and possible damage to the plankton community (Srain et al., 2021). Plankton is a good place for PPCPs to bioaccumulate and biomagnify their way up the food chain (Srain et al., 2021). This indicates that elevated PPCP concentrations in top predators might result from bigger species consuming the compounds that tiny plankton organisms acquire. Higher up the food chain animals may have more profound ecological consequences as a result of this (Kumar and Sharma, 2023). PPCP bioaccumulation in the environment gives rise to toxicity and impaired growth and development, which impact the organisms' overall well-being and ability to survive. The function and structure of aquatic habitats are also impacted, which has an impact on populations of various species and the overall health of the ecosystem (Kumar and Sharma, 2023).

2.4.3 Toxicity and Stress

The phenomenon of oxidative stress is generated by an imbalance between the creation and accumulation of reactive oxygen species (ROS) in cells (Okpoghono et al., 2018). Stressed phytoplankton cells produce reactive oxygen species (ROS) intracellularly when the environment is contaminated with PPCPs (Srain et al., 2021). ROS buildup affects phytoplankton cells (microalgae) by changing and inactivating the functions of biomolecules such as carbohydrates, lipids, proteins, and DNA. As a result, contact with PPCPs, algae, and cyanobacteria has been demonstrated to use enzymatic and non-enzymatic pathways to sequester and detoxify growing intracellular ROS concentrations (Pan et al., 2018). The existence of PPCPs causes oxidative stresses in phytoplankton, reducing their chances of survival. Aquatic organisms die as a result of acute toxicity (lethal effects) of PPCPs, whereas their chronic toxicity (sub-lethal effects) increases their capacity to disrupt aquatic organisms' homeostasis and interfere with endocrine systems, among other abnormalities (Srain et al., 2021).

2.4.4 Impacts on Biodiversity

Genetic, species, and ecosystem diversity are all included in biodiversity. The total biodiversity of aquatic ecosystems can be impacted by modifications to plankton communities. An ecosystem's total diversity of plankton species may be diminished if some species are especially vulnerable to PPCPs and experience a decline. The long-term viability of genetic diversity, populations, and ecosystems is compromised by PPCPs, which have negative effects on biodiversity at various scales. They may have detrimental effects on biodiversity because they can act on particular metabolic processes and cause a biological response at low concentrations (Néstor and Mariana, 2017).

2.5 INTERACTIONS BETWEEN TRANSGENERATIONAL AND ECOLOGICAL EFFECTS

Plankton populations exposed to PPCPs may experience changes in population size and structure as a result of transgenerational effects (He et al., 2023). For instance, if PPCPs disrupt the reproductive success of plankton, it may result in reduced population numbers over successive generations, affecting the availability of plankton as a source of food for higher trophic levels (He et al., 2023). Plankton communities may change as a result of transgenerational effects favoring specific plankton species more resistant to PPCPs. Changes in the composition of plankton can disturb the kinds of organisms that feed on plankton and compete with it for resources, which can impact the ecosystem as a whole (Gomaa et al., 2020). The aquatic ecosystem may grow more susceptible to other environmental stressors as the effects of PPCPs build up over generations. Multiple stressors working together can worsen the ecological effects and reduce the resilience of the ecosystem (Zhou et al., 2020). Transgenerational impacts can last for numerous generations, which infers that the ecological implications may last for a long time. This long-term impact might be especially problematic for aquatic ecosystems' general health and function. PPCP transgenerational impacts on plankton populations might increase the ecological implications of PPCP pollution by modifying the structure and function of plankton communities, impacting trophic relationships, and potentially lowering the stability and resilience of aquatic ecosystems (Kumar and Sharma, 2023).

2.6 MITIGATION AND MANAGEMENT APPROACH ON THE EFFECT OF PPCPs ON PLANKTON COMMUNITY

Pharmaceuticals and personal care products have persisted in natural waters because they are continuously added to aquatic environments and are not effectively removed. PPCPs can infiltrate aquatic environments via several routes, including leaching from landfills, agricultural field runoff, and wastewater treatment plant effluents (Gomaa et al., 2020). When introduced into aquatic environments, they harm aquatic organisms, and to address this problem, management techniques are required to lessen the pollution of PPCPs in the environment. Several methods are used to limit the incidence of PPCPs in aquatic environments. Generally speaking, there are three forms of remedy approaches for eliminating PPCPs: chemical, physical, and biological (Loganathan et al., 2023). Advanced oxidation processes (AOPs) are chemical techniques that are frequently used in tertiary treatments. Membrane separation and adsorption are the most often utilized physical procedures (Loganathan et al., 2023). Microbial degradation in the activated sludge process is the primary biological process and secondary treatment technique in wastewater treatment plants (WWTPs) (Roman et al., 2022). Microbial degradation has several benefits including affordable costs and comfortable working conditions (Loganathan et al., 2023). Ecologically, it is also an important way to solve the soil pollution caused by PPCPs, and it is very effective (Zeng et al., 2022). Limiting their occurrence in plankton communities can be achieved by improving wastewater treatment processes to eliminate or significantly lower PPCP concentrations before their release into the environment (Loganathan et al., 2023). An indispensable constituent in lowering PPCP pollution in the environment is PPCP use regulation which involves putting limitations on the use of specific chemicals. Proper disposal of PPCPs is also necessary for decreasing pollution in the environment. Limiting the persistence of pharmaceuticals and personal care products in aquatic environments can also be achieved through green chemistry by using less toxic or non-toxic chemicals during their production (Kumar and Sharma, 2023).

2.7 CONCLUSION

Pharmaceuticals and personal care products which are emergent pollutants in aquatic environments possess ecological effects by changing the composition and role of plankton populations in addition to transgenerational effects. This may cause disturbances that affect multiple generations of plankton, with a focus on interferences with physiology, behavior, growth, and reproduction. The fact that plankton are vulnerable to persistent pollution throughout their life cycle, there is emerging evidence that PPCP occurrence in aquatic environments has a deleterious impact on plankton. PPCP transgenerational impacts on plankton populations can worsen the ecological implications of PPCP pollution by modifying the role and setup of plankton communities, impacting trophic relationships, and potentially lowering the stability and resilience of aquatic ecosystems.

2.8 RECOMMENDATION

The complex collaboration between transgenerational and ecological effects highlights how important it is to understand how PPCPs may upset the aquatic ecosystem equilibrium. Therefore, it is critical to preserve aquatic ecosystems and the population of plankton. As a result, to safeguard the general health of aquatic ecosystems, the influence of PPCPs on plankton populations must be limited.

2.9 FUTURE OUTLOOK

To understand the inner mechanism of plankton's response to PPCP toxicity, more research is required. Certain endpoints, like genetic damage and the metabolic pathway, haven't been fully investigated, though. To understand the harmful effects of cosmetic wastes and PPCP on plankton communities, more research should be done on metabolic, enzymatic, or cell signaling pathways. The mechanisms of molecular evolution and the synthesis of active metabolites may be revealed by studying the metabolism of plankton, specifically the fatty acid and carbohydrate metabolism pathways. Further research is also necessary on genotoxicity. Since compounds found in PPCPs possess the ability to harm genetic information, it may pose a risk to future generations of the species through reproduction and potentially impact plankton populations through gene transfer.

REFERENCES

Baho, D.L., Pomati, F., Leu, E., Hessen, D.O., Moe, S.J., Norberg, J., & Nizzetto, L. (2019). A single pulse of diffuse contaminants alters the size distribution of natural phytoplankton communities. *Science of the Total Environment*, 683, 578–588.

Bickham, J. (2011). The four cornerstones of Evolutionary Toxicology. *Ecotoxicology*, 20(3), 497–502.

Brodin, T., Piovano, S., Fick, J., Klaminder, J., Heynen, M., & Jonsson, M. (2014). Ecological effects of pharmaceuticals in aquatic systems—impacts through behavioral alterations. *Philosophical Transactions of the Royal Society B: Biological Sciences*, 369(1656), 20130580.

Chen, L., Lam, J.C., Hu, C., Tsui, M.M., Lam, P.K., & Zhou, B. (2019). Perfluorobutanesulfonate exposure skews sex ratio in fish and transgenerationally impairs reproduction. *Environmental Science & Technology*, 53, 8389–8397.

Chia, M.A., Lorenzi, A.S., Ameh, I., Dauda, S., Cordeiro-Araújo, M.K., Agee, J.T., Okpanachi, I.Y., & Adesalu, A.T. (2021). Susceptibility of phytoplankton to the increasing presence of active pharmaceutical ingredients (APIs) in the aquatic environment: A review. *Aquatic Toxicology*, 234, 105809.

Ebele, A.J., Abou-Elwafa, A.M., & Harrad, S. (2017). Pharmaceuticals and personal care products (PPCPs) in the freshwater aquatic environment. *Emerging Contaminants*, 3(1), 1–16.

Ekta, B., & Pammi, G. (2018). Impact of antibiotics on plants. *International Journal of Pharmaceutical Sciences Review & Research*, 52, 49–33.

Gomaa, M., Zien-Elabdeen, A., Hifney, A.F., & Adam, M.S. (2020). Environmental risk analysis of pharmaceuticals on freshwater phytoplankton assemblage: Effects on alpha, beta, and taxonomic diversity. *Environmental Science & Pollution Research*, 28, 9954–9964.

Gomes, J., Costa, R., Quinta-Ferreira, R.M., & Martins, R.C. (2017). Application of ozonation for pharmaceuticals and personal care products removal from water. *Science of the Total Environment*, 586, 265–283.

He, Y., Zhang, Y., Zhou, W., Freitas, R., Zhang, Y., & Zhang, Y. (2023). Combined exposure of polystyrene microplastics and carbamazepine induced transgenerational effects on the reproduction of Daphnia magna. *Environmental Science & Pollution Research*, 30(25), 67596–67607.

Kallenborn, R., Brorström-Lundén, E., Reiersen, L.O., & Wilson, S. (2018). Pharmaceuticals and personal care products (PPCPs) in Arctic environments: Indicator contaminants for assessing local and remote anthropogenic sources in a pristine ecosystem in change. *Environmental Science & Pollution Research*, 25, 33001–33013.

Kayode-Afolayan, S.D., Ahuekwe, E.F., & Nwinyi, O.C. (2022). Impacts of pharmaceutical effluents on aquatic ecosystems. *Scientific African*, 17, e01288.

Kumar, D., & Sharma, S. (2023). Pharmaceutical and personal care product: A major threat to aquatic life. *Journal of Survey in Fisheries Sciences*, 10(4), 2791–2804.

Lee, I., Lee, J., Jung, D., Kim, S., & Choi, K. (2019). Two-generation exposure to 2-ethylhexyl 4- methoxycinnamate (EHMC) in Japanese medaka (Oryziaslatipes) and its reproduction and endocrine related effects. *Chemosphere*, 228, 478–484.

Loganathan, P., Vigneswaran, S., Kandasamy, J., Cuprys, A.K., Maletskyi, Z., & Ratnaweera, H. (2023). Treatment trends and combined methods in removing pharmaceuticals and personal care products from wastewater: A review. *Membranes (Basel)*, 13(2), 158.

Magureanu, M., Mandache, N.B., & Parvulescu, V.I. (2015). Degradation of pharmaceutical compounds in water by non-thermal plasma treatment. *Water Research*, 81, 124–136.

Maria, K. (2017). *Fate of Pharmaceuticals in the Environment—A Review Department of Ecology and Environmental Science (EMG).* UMEA University, Umeå, Sweden, 45–56.

Néstor, M.C., & Mariana, C. (2017). Impact of pharmaceutical waste on biodiversity. In Gómez-Oliván, L. (eds) *Ecopharmacovigilance. The Handbook of Environmental Chemistry*, vol. 66. Springer, Cham, 235–253.

Okpoghono, J., Achuba, F.I., & George, B.O. (2018). Influence of *Monodora myristica* extracts on lipid profile of rats fed crude petroleum oil contaminated catfish diet. *Sokoto Journal of Medical Laboratory Science*, 3(1), 19–27.

Pan, C.G., Peng, F.J., Shi, W.J., Hu, L.X., Wei, X.D., & Ying, G.G. (2018). Triclosan induced transcriptional and biochemical alterations in the freshwater green algae Chlamydomonas reinhardtii. *Ecotoxicology & Environmental Safety*, 148, 393–401.

Pomati, F., & Nizzetto, L. (2013). Assessing triclosan-induced ecological and transgenerational effects in natural phytoplankton communities: A trait-based field method. *Ecotoxicology*, 22, 779–794.

Rodea-Palomares, I., González-Pleiter, M., Martín-Betancor, K., Rosal, R., & Fernández-Piñas, F. (2015). Additivity and interactions in ecotoxicity of pollutant mixtures: Some patterns, conclusions, and open questions. *Toxics*, 3(4), 342–369.

Roman, M.D., Sava, C., Iluţiu-Varvara, D.A., Mare, R., Pruteanu, L.L., Pică, E.M., & Jäntschi, L. (2022). Biological activated sludge from wastewater treatment plant before and during the COVID-19 pandemic. *International Journal of Environmental Research & Public Health*, 19(18), 11323.

Romero-Murillo, P., Gallego, J.L., & Leignel, V. (2023). Marine pollution and advances in biomonitoring in cartagena bay in the Colombian Caribbean. *Toxics*, 11(7), 631.

Roveri, V., Guimarães, L.L., Toma, W., & Correia, A.T. (2022). Occurrence, ecological risk assessment and prioritization of pharmaceuticals and abuse drugs in estuarine waters along the São Paulo coast, Brazil. *Environmental Science & Pollution Research*, 29, 89712–89726.

Schaap, A., Dumon, J., & Den Toonder, J. (2016). Sorting algal cells by morphology in spiral microchannels using inertial microfluidics. *Microfluids & Nanofluids*, 20, 125.

Srain, H.S., Beazley, K.F., & Walker, T.R. (2021). Pharmaceuticals and personal care products and their sublethal and lethal effects in aquatic organisms. *Environmental Reviews*, 29(2), 142–181.

Tang, J., Wang, X., Yin, J., Han, Y., Yang, J., Lu, X., Xie, T., Akbar, S., Lyu, K., & Yang, Z. (2019). Molecular characterization of thioredoxin reductase in waterflea Daphnia magna and its expression regulation by polystyrene microplastics. *Aquatic Toxicology*, 208, 90–97.

Vaghela, K.B., Shukla, D.P., & Jain, N.K. (2023). A study of phytoplankton and zooplankton diversity in the River Sabarmati, Gujarat, India. *Asian Journal of Environment & Ecology*, 22(4), 28–38.

Vera-Chang, M.N., Moon, T.W., & Trudeau, V.L. (2019). Cortisol disruption and transgenerational alteration in the expression of stress-related genes in zebrafish larvae following fluoxetine exposure. *Toxicology & Applied Pharmacology*, 382, 114742.

Wang, M., Zhang, Y., & Guo, P. (2017). Effect of florfenicol and thiamphenicol exposure on the photosynthesis and antioxidant system of Microcystisflos-aquae. *Aquatic Toxicology*, 186, 67–76.

Wei, H., Tang, M., & Xu, X. (2023). Mechanism of uptake, accumulation, transport, metabolism and phytotoxic effects of pharmaceuticals and personal care products within plants: A review. *Science of the Total Environment*, 892, 164413.

Xin, X., Huang, G., & Zhang, B. (2020). Review of aquatic toxicity of pharmaceuticals and personal care products to algae. *Journal of Hazardous Materials*, 410(7), 124619.

Zeng, Y., Zhang, Y., Zhang, H., Wang, J., Lian, K., & Ai, L. (2022). Uptake and transport of different concentrations of PPCPs by vegetables. *International Journal of Environmental Research & Public Health*, 19(23), 15840.

Zhou, R., Lu, G., Yan, Z., Jiang, R., Bao, X., & Lu, P. (2020). A review of the influences of microplastics on toxicity and transgenerational effects of pharmaceutical and personal care products in aquatic environment. *Science of the Total Environment*, 732, 139222.

3 Analytical Study of the Environmental Effects of Emergent Pollutants on Plankton Community

Precious Onome Obiebi, Ebere Mary Eze, Solomon Ugochukwu Okom, Peter Mudiaga Etaware, Uduenevwo Francis Evuen, Odiri Ukolobi, Precious Emuoghenerue Omoruwou, Oke Aruoren, Irene Ebosereme Ainyanbhor, and Joshua Othuke Orogu

3.1 INTRODUCTION

Emerging pollutants (Eps) deposited in the environment are mostly natural substances and chemicals generated from pharmaceuticals, personal care products, pesticides, and industrial chemicals (Hena et al., 2020). These EPs are waste materials that are more hazardous and lethal to life simply because they cannot be curtailed by the regular laws governing global waste disposal or management. Commonly generated EPs may include a variety of compounds such as antibiotics, drugs, steroids, endocrine disruptors, hormones, industrial additives, chemicals, and microplastics (Hena et al., 2020). EPs are broadly classified into more than 121 types of unregulated chemicals that were detected in untreated water samples, by researchers, over time (Carmicheal, 1997). Due to their high level of resilience or recalcitrant nature, their transformation propensity, their capacity to form metabolites, and their ability to generate by-products, EPs can remain in the environment for longer periods (Abdulrazaq et al., 2021).

These chemicals and their by-products are known to cause severe negative effects on both terrestrial and aquatic environments, with the bulk of the damage ladened on the primary producers like phyto- and zooplankton and other microalgae used as food by marine animals (Luo et al., 2014). According to Xia et al. (2005), the pollution of the environment by these products is due to agrochemical leaching and inefficient treatment of domestic sewage before it is being released into water bodies like lagoons. The removal of emerging pollutants is usually not very efficient by simply adopting the use of conventional water treatment methods in developing

DOI: 10.1201/9781003362975-3

countries (Joyce et al., 2023). Therefore, this study was designed to investigate the harmful effects of EPs and how they can affect the normal functioning of the ecosystems if they are not properly managed. The secondary objective of this study was to ascertain the magnitude of production of EPs during the COVID-19 lockdown and, afterward, to effectively propose methods that can be used to curtail the production and/or indiscriminate disposal of EPs.

3.2 PLANKTON

Plankton is a term used to refer to very small tiny organisms in natural waters (marine and freshwater) that are mostly carried or drifted by tides and unable to swim well enough on their own either because they are too weak or too small to swim or, in some cases, non-motile (except for those that have flagellum) (Smith, 2013). Generally, the term plankton is collectively used for such organisms as algae, bacteria, protozoans, crustaceans, etc. Plankton is distinguished from nekton which is made up of strong swimming animals and from benthos which comprise burrowing and creeping organisms on the sea floor (Brierley, 2017). Plankton are mostly microscopic in size ranging from nanoplankton (2–20 μ) to mega plankton (20 cm). In natural water bodies, plankton are regarded as primary producers because they are the basis of the natural waters' food web. They are a very important source of food for many large and small organisms in the aquatic environment. They are usually easily affected by pollutants in the environment.

TABLE 3.1
Surface Water Organisms and the Impacts of Emergent Pollutants on Them.

S/n	Organism	Impacts of emergent pollutants on surface water organism	References
1.	Green algae	Destruction of chlorophyll and inhibition of photosynthesis affects metabolism, causes cell necrosis and eventual death	Sjollema et al. (2015)
2.	Blue-green algae (cyanobacteria)	Reduces their diversity and abundance	Joyce et al. (2023)
3.	Bacteria	Development of antibiotic-resistant species	Wang et al. (2021)
4.	Protozoa	Causes death to protozoa	Zhang et al. (2021)
5.	Dinoflagellates	Bioaccumulation in dinoflagellates resulting in biomagnification in higher trophic levels	Cole et al. (2015)
6.	Copepods	Bioaccumulation in copepods resulting in biomagnification in higher trophic levels	Cole et al. (2015)

According to Erondu and Solomon, 2017, the various surface water organisms that are affected by emerging pollutants and their effects on the organisms are stated in Table 3.1.

3.3 EMERGING POLLUTANTS: TYPES AND ENVIRONMENTAL EFFECTS

Stefanakis and Becker (2015) classified major emerging pollutants into two groups:

Group 1: Natural chemicals—includes agro-products, hormones, and steroids, as well as industrial additives and chemicals.
Group 2: Synthetic chemical—includes pharmaceuticals and personal care products.

3.3.1 PHARMACEUTICALS

Pharmaceuticals used for disease prevention and treatment in both humans and animals have been identified as common waste found in surface water and groundwater (Stefanakis and Becker, 2015). The pharmaceuticals which are considered emerging pollutants because of their recalcitrant nature in the ecosystem and the ability to cause adverse effects to aquatic organisms contain chemical compounds such as analgesics and anti-inflammatory drugs (ibuprofen, diclofenac, paracetamol, acetylsalicylic acid, etc.), psychiatric drugs (diazepam, carbamazepine, etc.), β-blockers (metoprolol, propranolol), lipid regulators (bezafibrate, clofibric acid, fenofibric acid), and X-ray contrasts (iopromide, iopamidol, diatrizoate) which are not broken down to their finite state which could be eco-friendly before the discharge into the environment (Stefanakis and Becker, 2015). Recent studies showed that a large number of pharmaceuticals and their metabolites were found at different concentrations (from ng/L to μg/L) in groundwaters that are used as drinking water sources in some remote areas of Germany and some medium-sized Mediterranean catchments (Rabiet et al., 2006; Teijon et al., 2010). Also, some classes of recalcitrant drugs were detected in effluents and water bodies in Sweden between 1988 and 1994 (Ekelund, 2015; Georgina, 2012).

It is common knowledge that some quantity of ingested pharmaceutical compounds, like synthetic hormones, anti-inflammatory, statins, antidepressants, antiepileptics, beta-blockers, contrast agents, antibiotics, etc., are largely excreted, either in their original form or as metabolites, mostly through urination after drug administration and partial metabolism by the body system. These excreted drugs (secondary wastes) can usually be found in urban wastewater, surface or runoff waters, and hospital sewers (Caliman and Gavrilescu, 2009; Yehya et al., 2015). Recalcitrant drugs discharged into the environment can also reach groundwater from water used for irrigation purposes. Antibiotics in the environment seem to spread in increasing quantities with a geometric increase in the diversity of the recalcitrant compounds. Roston (2019) in his scientific report stated that some rivers of the world have antibiotic concentrations up to 300 times higher than the recommended "safe" levels. According to the World Health Organization (WHO), the greatest threat to global food security, health, and development is microbial resistance to antibiotics. An increasing number of infections and diseases are difficult to treat because antibiotics used to treat them are increasingly becoming less effective due to environmental pollution (Cycon et al., 2019; Kraemer et al., 2019). These pollutants tend to reduce the population of plankton in the water body, thereby reducing the existing food source for marine animals that are dependent on them for survival.

3.3.2 Personal Care Products

The active ingredients in personal care products used in the production of cosmetics, toiletries, and perfumes constitute another group of emerging pollutants. Personal care products (PCP) are applied directly on the human skin and not for consumption (Daughton, 2005). The compounds used as fragrances (e.g. nitro, polycyclic and macrocyclic musk, phthalates) and the preservatives from shampoos, toiletries, and creams are of particular interest to biologist because of their harmful effects, particularly on aquatic lives. Compounds such as benzophenone from sunscreen lotions that can block UV light or such compounds as alkylate siloxane utilized in soaps, haircare products, lotions, etc., are also seen as emerging pollutants (Houtman, 2010). Also, disinfectants used in surface sterilization of organic or inorganic surfaces are known for their biocidal and antiseptic effect (e.g. triclocarban, triclosan, chlorophane, etc.). Triclosan is used in a wide variety of consumer products like toothpaste, hand soap (personal care products), toys, and socks (Petrovic et al., 2003). Perfluorinated compounds, such as perfluoroctane sulfonates, and perfluoroctanoic acid are examples of recalcitrant compounds with special chemical properties mostly used as impermeable coatings to water, dirt, grease in the spray forms for leather, textiles, as well as in cookware nonstick PTFE (Apetroaei et al., 2020). These perfluorinated compounds are mostly toxic chemicals that can gradually accumulate in the human body leading to the formation of carcinogenic, nephrotoxic, hepatotoxic, bronchitis, or cardiovascular infections (Skutlarek et al., 2006). Also, the accumulation of these harmful substances in water bodies can cause a massive reduction in the plankton community resulting in food shortage for aquatic animals.

3.3.3 Agro-Products

The widespread use of agro-products in the environment, mostly in the form of herbicides, insecticides, fungicides, plant growth regulators, and bactericides, is a threat to surface water and groundwater supplies. Their unregulated widespread use for agricultural purposes is an important source of contamination to the aquatic environment (Houtman, 2010). DDT, chlordane, and aldrin, which are highly persistent chemical compounds, have been reported to hurt the environment with particular reference to birds and fishes (Houtman, 2010). This is due to their bioaccumulation potential (Houtman, 2010). Kuster et al. (2009) reported that glyphosates, organophosphorus herbicides, triazines, thiocarbamates, and chlorophenoxy acetic acids have been grouped as polar pesticides that are less persistent in the environment. These pesticides can actively impede the growth of the plankton community in the nearby water body into which they are leached, thereby causing massive death of plankton (D'Costa et al., 2017).

3.3.4 Hormones and Steroids

Hormones and steroids such as sex hormones, mostly estrogens (e.g. progesterone, oestriol, and estrone) and androgens (e.g. testosterone and androstenedione), were found in human fecal samples (Stefanakis and Becker, 2015). Also, phytoestrogens and plant sterols were among the noted steroids found in excreted wastes from the human body too. These leachates found in small quantities in surface water may not influence the population of the plankton community and, as such, may not be a major problem at the moment, but shortly, this might become a serious problem (Stefanakis and Becker, 2015). Therefore, solutions regarding its management and/or control should be proffered now.

3.3.5 Industrial Additives and Chemicals

Industrial additives and chemical agents comprise many organic and inorganic compounds that have been identified in aquatic ecosystems and classified as emerging pollutants. This class of EPs includes complex agents, or substances used in industrial processes and production, particularly in the chemical industry. An example of such complexing agents that are used in common products such as soap and toothpaste is ethylenediaminetetraacetic acid (EDTA), benzotriazoles used as anti-corrosion agents for metal coatings to protect metals in contact with fluids (e.g. in engine coolants or anti-freezing liquids) (Apetroaei et al., 2020). These pollutants are soluble in water, non-biodegradable, recalcitrant, and poorly removed in wastewater treatment plants (Houtman, 2010). Gasoline additives such as methyl tertiary butyl ether (MTBE) and ethyl tertiary butyl ether (ETBE) are also considered as types of emerging pollutants. They are used to optimize combustion, reduce emissions, and help prevent engine knocking (Stanciu and Constantin, 2018). However, MTBE and ETBE have some adverse effects like quick dispersion into the environment due to their high volatility and solubility rates and are also non-biodegradable. These substances can increase or lower the acidity or alkalinity level of stagnant water, thereby altering the population of the plankton community in the environment.

3.4 IMPACT OF EMERGENT POLLUTANTS ON SURFACE WATER ORGANISMS

The plankton community occupies a vital position in the food chain of any aquatic ecosystem such that every organism living in water depends either directly or indirectly on plankton for survival. They constitute the primary producers and account for up to 50% of the primary production of food on earth. Plankton are also used frequently as bioindicators in monitoring changes in the ecology of aquatic systems (Paul et al., 2016). They are, therefore, management tools for supervising the quality of the ecological system and indicators for necessary actions. An example is the prevention of algal bloom and toxic contamination from unidentified sources. This is because plankton can react to slight variations in abiotic factors such as pollutants, light, temperature, heavy metals, pH, nutrient concentration, and salinity as well as biotic factors such as parasites and predators (Wiltshire et al., 2015).

Despite the obvious benefits of phytoplankton to the aquatic lives and, by extension, all life forms, emerging pollutants can influence phytoplankton and other water surface microorganisms by inducing alterations in composition and abundance of species and, in extreme cases, mortality of very sensitive ones (Corcoll et al., 2015). Chia et al. (2021) reported that exposing the phytoplankton community to pharmaceutical products results in growth inhibition, reduced cell sizes, cell deformities, inhibition of photosynthesis, and general biomass reduction. These changes in their physiology affect their photosynthetic activities, contribution to nutrient cycling, ecological balance, and energy transfer in the aquatic environment (Liu et al., 2011; Smith, 2013). In addition, phytoplankton is faced with multiple active pharmaceutical ingredients (APIs) in the natural ecosystem which may have an antagonistic, synergistic, or additive effect on them (Chia et al., 2021).

3.5 MITIGATION OF EMERGENT POLLUTANTS

Due to the increasing rate of discharge of emerging pollutants on water bodies and their adverse effects on aquatic lives and life forms generally, it is imperative to discuss the various technologies that are currently in use for the mitigation of these pollutants.

3.5.1 Adsorption Technology

Adsorption as a process involves the transfer of materials between two phases that is either liquid-liquid, liquid-solid, and gas-liquid or gas-solid interface (DeGisi et al., 2016). In this technology, adsorbents (i.e. bamboo dust, silica gel, activated carbon, fly ash, peat, etc.) are employed to adsorb pollutants of concern from wastewater using intermolecular forces (DeGisi et al., 2016; Yaashikaa et al., 2016). This technology is capable of very specific removal of the emergent pollutants by accumulation of the toxins in the adsorbents (DeGisi et al., 2016).

3.5.2 Membrane Filtration

Membrane filtration technology is a physical method in which membranes produced from materials having properties of filtration such as pore size, surface charge, and hydrophobicity are used to filter contaminants (Nghiem and Fujioka, 2016). This technology is categorized into ultrafiltration (UF), nanofiltration (NF), microfiltration (MF), forward osmosis (FO), and reverse osmosis (RO) This categorization is dependent on the pore sizes of the membrane filter. These processes can reduce emerging pollutants up to as much as 99% if well carried out (Nghiem and Fujioka, 2016). Membrane filtration technology is effective in removing pharmaceuticals (i.e. anti-inflammatory drugs, fluoroquinolones, and sulfonamide) and hormones (i.e. testosterone, progesterone, and estradiol) (Nghiem and Fujioka, 2016).

3.5.3 Coagulation-Flocculation

Coagulation is the process of dispersing dye solutions to form flocs. Flocculation is the joining of aggregated flocs to form larger flocks and agglomerates which settle to the bottom due to the force of gravity (Teh et al., 2016). Coagulation-flocculation is a process considered very effective for thc removal of large suspended particles of dyes dispersed in wastewater resulting in its coloration (Teh et al., 2016). During the process, coagulants such as ferric sulfate ($Fe_2(SO4)_3{\cdot}7H_2O$), ferric chloride ($FeCl_3{\cdot}7H_2O$), aluminum sulfate ($Al_2(SO4)_3{\cdot}18H_2O$), and lime ($Ca(OH)_2$) combine with the pollutants to remove them by sorption. Pharmaceuticals such as hydrochlorothiazide and warfarin have been reported by Mohamed et al. (2021) to be effectively removed using aluminum sulfate ($Al_2(SO4)_3{\cdot}$).

3.5.4 Solvent Extraction

This technique is widely employed in the elimination of inorganic and organic pollutants in water bodies. In this method, the solute particles are first transferred to a solvent from water. Thereafter, the solute is separated from the solvent, and lastly, the solvent is recovered. The removal of phenols and cresols from polluted water is very efficient using the solvent extraction method (Rajan et al., 2016).

3.5.4.1 Advanced Oxidation Processes (AOPs)

Advanced oxidation processes (AOPs) are mitigation procedures that are based on the oxidation of emergent pollutants by generating hydroxyl (OH) radicals. They include ozonation (O_3), electrochemical oxidation, hydrogen peroxide (H_2O_2), UV light, and photocatalysis process. Most times, UV irradiations and ozone are deployed for the efficient removal of emergent pollutants. AOPs are very efficient in removing non-biodegradable pollutants like pesticides and volatile organic compounds (VOCs). AOPs can be used to remove many organic compounds at the same time without the production of hazardous substances in water (Deng and Zhao, 2015)

3.6 THE FUTURE OF EMERGENT POLLUTANTS

The rapidly growing population of the world is a recipe for the increasing demand for freshwater utilized for domestic and industrial activities. The increasing population also results in the demand and production of pharmaceuticals, personal body care products, hormones, etc., which eventually constitute emergent pollutants. Consequently, the generation of more wastewater cannot be evaded, thus making the management of emergent pollutants in water bodies a very challenging exercise. Currently, the treatment of wastewater poses a very serious challenge because of its effect on the environment and its dependence on socioeconomic circumstances. Several methods have been employed to eliminate/mitigate emergent pollutants. It is, however, difficult to identify the most suitable method to employ considering their ecological impacts, cost efficiency, feasibility, and operational obscurity; hence, a reasonable and effective mitigation exercise may require more than one method.

3.7 CONCLUSION

The increasing level of emergent pollutants in the environment has been proven to have adverse effects on the plankton community and food availability to animals in most aquatic habitats. The removal of emergent pollutants is more effective with the use of a combination of two or more of the following methods: adsorption, coagulation/flocculation, membrane filtration, solvent extraction, and advanced oxidation processes (AOPs) technologies as discussed earlier. Therefore, appropriate measures must be deployed in the mitigation of emergent pollutants to ensure an effective treatment of polluted water bodies as well as the treatment of raw wastes before disposal in water bodies. Moreover, efficient and effective eco-friendly bioremediation methods should be developed to remove existing emergent pollutants found in most water bodies around the world.

REFERENCES

Abdulrazaq, Y., Abdulsalam, A., Larayetan Rotimi, A., Aliyu Abdulbasit, A., et al. 2021. Classification, potential routes and risk of emerging pollutants/contaminant. *IntechOpen*. http://doi.org/10.5772/intechopen.94447.

Apetroaei, M. R., Avram, E. R., Atodiresei, D. V., and Apetroaei, G. M. 2020. Emerging pollutants—a potential threat to the marine environment. *Scientific Bulletin of Naval Academy* 23: 171–178.

Brierley, A. S. 2017. Plankton. *Current Biology* 27 (11): 478–483e.

Caliman, F. A., and Gavrilescu, M. 2009. Pharmaceuticals, personal care products and endocrine disrupting agents in the environment—A review. *Clean (Weinh)* 37: 277–303.

Carmicheal, W. W. 1997. The cyanotoxins. *Advances in Botanical Research* 27: 211–256. https://doi.org/10.1016/S0065-2296(08)60282-7

Chia, M. A., Lorenzi, A. S., Ameh, I., Dauda, S., Cordeiro-Araújo, M. K., Agee, T., Okpanachi, I. Y., and Adesalu, T. 2021. Susceptibility of phytoplankton to the increasing presence of active pharmaceutical ingredients (APIs) in the aquatic environment: A review. *Aquatic Toxicology* 234: 105809.

Cole, M., Lindeque, P., Fileman, E., Helsband, C., and Galloway, T. S. 2015. The impact of polystyrene microplastics on feeding, function and fecundity in the marine copebud *Calanus helgolandicus*. *Environmental Science & Technology* 49: 1130–1137.

Corcoll, N., Casella, M., Huerta, B., Guasch, H., Acuña, V., Rodríguez, M. S., Barceló, S. C. A., and Salbater, S. 2015. Efects of flow intermittency and pharmaceuticam exposure on the structure and metabolismo of stream bioflms. *Science Total Environment* 503: 159–170.

Cycon, M., Mrozik, A., and Piotrowska-Sege, Z. 2019. Antibiotics in the soil environment—Degradation and their impact on microbial activity and diversity. *Frontiers in Microbiology*. https://doi.org/10.3389/fmicb.2019.00338.

Daughton, C. G. 2005. Emerging chemicals as pollutants in the environment: A 21st century perspective. *Renewable Resources Journal* 23(4): 6–23.

DeGisi, S., Lofrano, G., Grassi, M., and Notarnicola, M. 2016. Characteristics and adsorption capacities of low-cost sorbents for wastewater treatment: A review. *Sustainable Materials and Technology* 9: 10–40.

Deng, Y., and Zhao, R. 2015. Advanced oxidation processes (AOPs) in wastewater treatment. *Current Pollution Reports* 1(3): 167–176. http://doi.org/10.1007/s40726-015-0015-z

Ekelund, N.2015. Fate of four pharmaceuticals in aquatic ecosystems: Investigating the role of U.V-light and animal assimilation as dissipation factors analysis of illicit drugs in wastewater and surface water. *Mass Spectrometry Reviews* 27: 378–394.

Erondu, C. J., and Solomon, R. J. 2017. Identification of plankton (zooplankton and phytoplankton) behind girls' hostel, University of Abuja, Nigeria. *Direct Research Journal of Public Health and Environmental Technology* 2(3): 21–29.

Georgina, C. 2012. Metabolism and excretion: Eliminating drugs from the body. *Nurses New Zealand* 18: 20–24.

Hena, S., Gutierrez, L., and Croué, J. P. 2020. Removal of pharmaceutical and personal care products (PPCPs) from wastewater using microalgae: A review. *Journal of Hazardous Materials*. https://doi.org/10.1016/j. jhazmat.2020.124041.

Houtman, C. J. 2010. Emerging contaminants in surface waters and their relevance for the production of drinking water in Europe. *Journal of Integrative Environment Science* 7(4): 271–295. http://doi.org/10.1080/1943815X.2010.511648.

Joyce, A., Anny, K., Priscilla, C., Jascieli, C., and Indianara, C. 2023. Emerging contaminants in the aquatic environment: Phytoplankton structure in the presence of sulfamethoxazole and diclofenac. *Environmental Science and Pollution Research* 30: 46604–46617.

Kraemer, S. A., Ramachandran, A., and Perron, G. G. 2019. Antibiotic pollution in the environment: From microbial ecology to public policy. *Microorganisms* 7: 180.

Kuster, M., de Lopez, A. M., and Barcelo, D. 2009. Liquid chromatography-tandem mass spectrometric analysis and regulatory issues of polar pesticides in natural and treated waters. *Journal of Chromatography*: 520–529.

Liu, B., Nie, X., Liu, W., Snoeijs, P., Guan, C., and Tsui, M. T. K. 2011. Toxic effects of erythromycin, ciprofloxacin and sulfamethozazole on photosynthetic apparatus in *Selenastrium capricornutum*. *Ecotoxicology and Environmental Safety* 74(4): 1027–1035.

Luo, Y., Guo, W., Ngo, H. H., Nghiem, L. D., Hai, F. I., Zhang, J., Liang, S., and Wang, X. C. 2014. A review on the occurrence of micropollutants in the aquatic environment and their fate and removal during wastewater treatment. *Science of the Total Environment* 473: 619–641.

Mohamed, N. M. H., Wong, S., Ngadi, N., Mohammed, I. I., and Opotu, L. A. 2021. Assessing the effectiveness of magnetic nanoparticles coagulation/flocculation in water treatment: A systematic literature review. *International journal of Environmental Science and Technology* 28(22): 1–22. http://doi.org/10.1007/s13762-021-03369-0.

Nghiem, L. D., and Fujioka, T. 2016. Removal of emerging contaminants for water reuse by membrane technology. In: *Emerging Membrane Technology for suitable Water Treatment*. Amsterdam, Netherlands, 217–247. http://doi.org/10:1016/B978-0-444-63312-5.00009-7.

Paul, S., Wooldridge, T., and Perissinotto, R. 2016. Evaluation of Abiotic Stresses of temperate estuaries by using resident zooplankton: A community vs population approach. *Estuarine Coastal Shelf Science* 170: 102–111.

Petrovic, M., Gonzalez, S., and Barcelo, D. 2003. Analysis and removal of emerging contaminants in wastewater and drinking water. *TrAC Trends in Analytical Chemistry* 22: 685–696.

Rabiet, M., Togola, A., Brissaud, F., et al. 2006. Consequences of treated water recycling as regards pharmaceuticals and drugs in surface and ground waters of a medium-sized Mediterranean catchment. *Environmental Science & Technology* 40(17): 5282–5288.

Rajan, A., Sreedharan, S., and Babu, V. 2016. Solvent extraction and adsorption technique for the treatment of pesticide effluent. *Civil Engineering and Urban Planning: An International Journal* 3(2): 155–165. http://doi.org/10.5121/civej.2016.3214.

Roston, B. A. 2019. Antibiotics found in rivers at up. *SlashGear*. Available online: Accessed on 15th July 2020. https://www.slashgear.com/antibioticsfound-in-rivers-at-up-to-300-times-over-safe-levels-27578126/.

Sjollema, S. B., Redondo-Hasselerharm, P., Leslie, H. A., and Kraak, M. H. S. 2015. Do plastic particles affect microalgae photosynthesis and growth? *Aquatic Toxicology* 170: 259–261.

Skutlarek, D., Exner, M., and Farber, H. 2006. Perfluorinated surfactants in surface and drinking water. *Environmental Science and Pollution Research* 13: 299–307.

Smith, D. J. 2013. Aeroplankton and the need for a global monitoring network. *BioScience* 63(7): 515–516. http://doi.org/1525/bio.2013.63.7.3.

Stanciu, T., and Constantin, A. 2018. Theoretical and experimental study of turbulent gas flow through a pneumatic mechanism. *International Multidisciplinary Scientific GeoConference Surveying Geology and Mining Ecology Management SGEM* 18(3.2): 1205–1212.

Stefanakis, A., and Becker, J. A. 2015. A review of emerging contaminants in water: Classification, sources and potential risks. In: *Impact of Water Pollution on Human Health and Environmental Sustainability*, 1st ed. Edited by McKeown, E., and Bugyi, G. IGI Global, Chapter 3, 55–80. http://doi.org/10.4018/978-1-4666-9559-7.ch003.

Teh, C. Y., Budiman, P. M., Shak, K. P. Y., and Wu, T. Y., 2016. Recent advancement of coagulation–flocculation and its application in wastewater treatment. *Industrial & Engineering Chemistry Research* 55(16): 4363–4389. http://doi.org/10.1021/acs.iecr.5b04703.

Teijon, G., Candela, L., Tamoh, K., et al. 2010. Occurrence of emerging contaminants, priority substances (2008/105/CE) and heavy metals in treated wastewater and groundwater at Depurbaix facility (Barcelona, Spain). *Science of the Total Environment* 408(17): 3584–3595.

Wang, Y., Lu, J., Zhang, S., Li, J., Mao, L., Yuan, Z., Bond, P. L., and Guo, J. 2021. Non-antibiotic pharmaceuticals promote the transmission of multidrug resistance plasmids through intra and intergenera conjugation. *The ISME Journal*, 15(9): 2493–2508. http://doi.org/10.1038/s41396-021-00945-7.

Wiltshire, K. H., Boersma, M., Carstens, K., Kraberg, A. C., Peters, S., and Scharfe, M. 2015. Control of phytoplankton in a shelf sea: Determination of the main divers based on the Helgoland Roads. *Times Series Journal Sea Resources* 15: 42–52.

Xia, K., Bhandari, A., Das, K., and Pillar, G. 2005. Occurrence and fate of pharmaceuticals and personal care products (PPCPs) in biosolids. *Journal of Environmental Quality* 34(1): 91.

Yaashikaa, P. R., Kumar, P. S., Varjini, S. J., and Saravanan, A. 2016. Advances in production and application of biochar from lignocellulosic feedstocks for remediation of environmental pollutants. *Bioresource Technology* 292: 122030.

Yehya, T., Favier, L., Kadmi, Y., Audonnet, F., Fayad, N., Gavrilescu, M., and Vial, C. 2015. Removal of carbamazepine by electrocoagulation: Investigation of some key operational parameters. *Environmental Engineering and Management Journal* 14: 639–645.

Zhang, X., Wang, J., Dua, B., Jiang, S. 2021. Degradation of sulfamethoxazole in water by a combined system of ultrasound/PW12/KI H202. *Separation and Purification Technology* 270: 118790. http://doi.org/10.106/j.seppur.2021.118790.

4 Role of Emergent Pollutants on the Macro-Community of Planktons

Ecosystem Influence

Joel Okpoghono, Ovigueroye Ekayoda,
Osikemekha Anthony Anani,
Fegor Endurance Isoje, Ebere Mary Eze,
Solomon Ugochukwu Okom,
Ochuko Ernest Jessa, Bishop Uche Okobia, and
Joshua Othuke Orogu

4.1 INTRODUCTION

Emerging pollutants (EPs) have been given the designation of "emerging" due to the increasing level of worry associated with them (Egbuna et al., 2021). Industrial, municipal, and domestic wastewater are the main sources of their widespread spread in the aquatic environment, and there is an inextricable relationship between these contaminants and wastewater. A wide range of substances are considered emerging pollutants, including chemicals, pharmaceuticals, industrial additives, pesticides, personal care items, microbeads, and microplastics (Krishnakumar et al., 2022). The widespread recognition of the contamination of aquatic ecosystems by these emerging pollutants has led to a significant public health concern, as these pollutants may pose a threat to the organisms that make up the aquatic habitat. EPs are directly released into rivers from industrial wastewater treatment facilities, raising concerns about their potential environmental fate (sorption at the sediment, degradation, or transport in the aqueous phase) (Geissen et al., 2015).

Phytoplankton use the sun's energy to create their nourishment from the nutrients in the water (Batten et al., 2019; Lobus and Kulikovskiy, 2023). Zooplankton acts as a link in the food chain, delivering energy from planktonic algae (primary producers) to bigger invertebrate predators and fish that eat them. Rapid changes in the environment can cause plankton communities to shift, which can lead to a variety of issues in the food chain (Lomartire et al., 2021). The sensitivity of zooplankton to alterations in aquatic environments is high. Differences in species composition, abundance, and distribution of body size can be used to identify the effects of environmental disturbances (Cavan et al., 2017).

DOI: 10.1201/9781003362975-4

EPs have the potential to alter species composition and abundance in aquatic environments, leading to competition between more resilient species and the extinction of the most vulnerable ones. The presence of EPs in the aquatic ecosystem has a deleterious influence on phytoplankton (algae) by inhibiting growth, causing cell abnormalities and lower cell sizes, inhibiting photosynthesis, and reducing biomass (Chia et al., 2021). Emerging contaminants have the potential to contaminate aquatic habitats and it is crucial to highlight that they are not recently produced substances, and they may have been present in the environment for a long time (Duarte et al., 2023). Research on EPs interactions with the aquatic biota particularly those aiming to figure out the relationship between emerging contaminants and plankton communities is scarce and is in its early stages. Therefore, this book chapter aims to review the environmental effects of emergent pollutants on the plankton community.

4.2 TYPES OF EMERGENT POLLUTANTS

4.2.1 Pharmaceuticals and Personal Care Products (PPCPs)

These are a broad category of chemicals that includes all pharmaceuticals (prescription and over-the-counter medications) that are largely used to prevent or cure diseases or sickness, whereas personal care products are non-medicinal consumer chemicals, which include moisturizers, perfumes (musks) in lotions and soaps, lipsticks, shampoos, hair colors, UV filters in sunscreens, deodorants, and toothpaste, are intended to improve the quality of everyday living (Osuoha et al., 2023). PPCPs are one of the most numerous EPs types (Liu et al., 2021). They are persistent in aquatic ecosystems due to their continuous release from treated and untreated wastewater discharges.

Recent studies have shown that PPCPs induce changes in phenotypic diversity, community composition, and community metabolism of phytoplankton assemblages, thus supporting the claim that PPCPs should be considered as an environmental stressor of primary concern. Aquatic toxicity of various PPCPs (such as erythromycin [antibiotic], fluoxetine [antidepressant], naproxen [a nonsteroidal anti-inflammatory], and gemfibrozil [lipid regulator]) on some isolated plankton species had been carried out (El-Bassat et al., 2011). The findings showed that algae (*Chlorella vulgaris* and *Ankestrodesmus falcatus*), protozoa (*Paramecium caudatum*), rotifers (*Brachionus calyciflorus*), and cladocerans (*Daphnia longispina* showed reduction in their defensive mechanisms that lead to decrease in antioxidant enzymes: superoxide dismutase (SOD) and catalase (CAT) activities and elevation in lipid peroxidation (TBARS) levels after exposure to pharmaceuticals products. The results suggest that the pharmaceutical products investigated have toxic effects on the aquatic organisms which may cause species abundances to fluctuate randomly, lowering diversity within communities (El-Bassat et al., 2011).

The impacts of these chemicals may be modest since PPCPs are commonly found in low quantities in the environment (Liu et al., 2021). PPCPs are so common and have ecotoxicological effects on the ecosystem with the growing environmental crisis. Since they are present in aquatic systems all over the world, PPCPs are contaminants of global concern. Industrial and household processes discharge these chemicals into aquatic environments regularly. Pharmaceutical industries, medical laboratories, affluent, household processes, and treated water discharge PPCP (medications such as analgesics, antibiotics, hormones in various forms, anti-diabetics, antihypertensives, and a variety of other health-related substances) into aquatic environments regularly which may get their way into groundwater. This could lead to contamination of water reservoirs (Narayanan et al., 2022).

4.2.2 Industrial Chemicals

Industrial chemicals, crude petroleum oil (Okpoghono et al., 2021), and many diverse synthetic organic compounds such as flame retardants, polychlorinated alkanes, plasticizers, bisphenol A, chlorinated solvents, perfluoro alkylated compounds, etc., are utilized in vast amounts all over the world for a variety of reasons (Egbuna et al., 2021). Numerous chemicals are utilized in the chemical industry as intermediates (plasticizers, dyes, resins) (Emegha et al., 2023) or as food additives, antioxidants surfactants, and detergents. As a result, these compounds are constantly discharged into the environment, mostly as a result of wastewater from industries and households (Calvo-Flores et al., 2017). The aquatic environment can be continuously exposed to complex mixtures of industrial chemicals and their bioactive metabolites through various routes. It is now commonly acknowledged that these chemicals can contaminate surface waters (Calvo-Flores et al., 2017). This has grown to be a significant public health concern since there is a chance that it will damage the acknowledged freshwater resource. Table 4.1 presents the environmental effects of industrial EPs.

TABLE 4.1
Effects of Industrial EPs.

	Industrial EPs	Effects
i	Flame retardants	Decrease in the number of large zooplankton and the amount of food available to planktivorous fish (Osier and Cohen, 2020).
ii	Polychlorinated alkanes	Lower the marine phytoplankton's rate of cell division, which will indirectly lower the total amount of carbon fixed by photosynthetic processes (Roy and Martin, 2014).
iii	Plasticizers	Restricts plankton's capacity to reproduce because of their extreme fragmentation and resilience to deterioration (Egbuna et al., 2021).
iv	Bisphenol A	Decreases *Chlorella pyrenoidosa* cell proliferation, chlorophyll concentration, and oxidative stress (Li et al., 2022).
v	Chlorinate solvents	Reduce the primary productivity, intracellular esterase activity, and chlorophyll auto fluorescence in chlorella cells (Ebenezer et al., 2012).
iv	Perfluoroalkylated compounds	Decreases chlorophyll a content in microalgae (Mojiri et al., 2023).

4.2.3 Microplastics

Over the years, public awareness of microplastic (MPs) contamination has grown dramatically. MPs polymers have been discovered in a variety of environments, including water, land, and the atmosphere. Polyethylene (PE) and polystyrene (PS) are the most prevalent MPs, followed by polypropylene (PP), polyvinylchloride (PVC), polyethylene terephthalate (PET), polyamide (PA), and polyvinyl alcohol (PVA) (Arienzo et al., 2021). MPs enter the environment by direct discharge of plastic garbage into bodies of water as well as discharge from industrial and household wastewater treatment facilities (WWTPs). MPs have the potential to not only be a hazardous pollutant but also to carry and transport additional substances of concern through the ecosystem. They accumulate and persist in environments where they can interact with organisms for extended periods because of their slow rate of degradation (Wu et al., 2019). The chemical makeup of the primary organic pollutants found in microplastics is shown in Figure 4.1. MPs have been observed to coexist with a variety of environmental contaminants and to serve as vectors for the transfer of these chemical contaminants to aquatic and terrestrial organisms from soil and water (Arienzo et al., 2021). Compounds such as DDT, PCBs, DDE, phenanthrene, bisphenol A, and PBDEs are mainly found in microplastics and are constantly discharged into the environment, regularly as a result of wastewater from industries and households.

4.2.4 Pesticides

The extensive usage of dangerous chemical-based insecticides has a bigger influence on the environment (Rajmohan et al., 2020). Agricultural activities are the principal sources of pesticide contamination in the environment (Pullagurala et al., 2018).

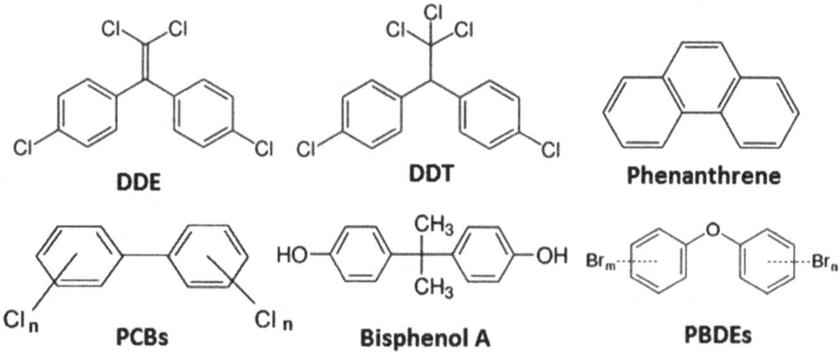

FIGURE 4.1 Chemical composition of the primary biological contaminants found in microplastics.

Note: DDT = dichlorodiphenyltrichloroethane; PCBs = polychlorinated biphenyls; DDE = 1,1-dichloro-2,2′-bis-chlorophenyl-ethylene; PBDEs = polybrominated diphenyl ethers.

Source: Arienzo et al. (2021).

Among the highly persistent substances found in agricultural soil are pesticides, and using them excessively can cause pollutants to build up in the environment. These pollutants put the ecosystem at higher risk and are toxic to humans (Pullagurala et al., 2018). Using spray drifting, aerial spraying, industrial and domestic sewage discharge, negligent disposal of empty chemical containers, and equipment washing, the hazardous chemical pesticides are released into aquatic environments. Pesticide-contaminated water poses a significant hazard to aquatic life, causing significant alterations in aquatic creature physiology (Mahmood et al., 2016).

4.3 PATHS OF EMERGING POLLUTANT INTO THE AQUATIC ENVIRONMENT

A few of the primary routes via which emerging pollutants enter aquatic environments are agricultural runoff, land-based sources, atmospheric deposition, direct contamination, and wastewater discharges from households (Arienzo et al., 2021). These pathways have the potential to have severe ecological and human health consequences.

The environment is exposed to trillions of tons of chemically active material as a result of agricultural practices and these chemicals have been reported to be transported by runoff (i.e. overland flow) in agricultural operations (Mishra et al., 2023). Agricultural chemicals which are pesticides (fertilizers, insecticides, and herbicides) are used in agricultural operations, and during rainy events, these pollutants can be carried into bodies of water via runoff (Mishra et al., 2023). Agricultural chemicals are widely dispersed and are transported by runoff, erosion, or leaching until they reach the aquatic environment. Agricultural chemicals metabolites are quickly absorbed by the soil and the sorption behavior of the compounds, the presence of fertilizer in the soil matrix, and the character of the land to which the fertilizer is applied all impact the movement of these molecules into the aquatic ecosystem (Mishra et al., 2023). Land management practices appear to have a major influence on the transportation of EPs to surface waters. Atmospheric deposition is a route in which EPs can be deposited in the ecosystem (Egbuna et al., 2021). Certain EPs, like chemicals, heavy metals, and airborne microplastics, can be carried by air currents and end up in water bodies. This usually occurs in places with a high concentration of industrial activity. EPs in the atmosphere may include volatile organic solvents, various particles, and bioaerosols (Egbuna et al., 2021). Through naturally occurring processes of deposition, atmospheric elements such as ozone, nitrogen- and sulfur-containing compounds, and particulate matter can make their way into soil, marine, and freshwater bodies. These atmospheric gases can combine with suspended water in the atmosphere and be carried away by rain, snow, or fog. Bioaerosols, different particles, and volatile organic solvents are examples of EPs that can be found in the atmosphere. Airborne components like ozone, compounds containing sulfur and nitrogen, and particulate matter can enter freshwater, marine, and soil through naturally occurring processes of deposition. These atmospheric gases can mix with atmospheric suspended water, which can then be delivered by rain, snow, or fog to humans (Saquib et al., 2021).

EPs direct contamination of the aquatic community can occur by accidental spills or leaching and seepage. Contaminants seeping from landfills and waste disposal facilities can contaminate groundwater. These contaminants may ultimately find their way into surrounding bodies of surface water (Mishra et al., 2023). EPs enter the aquatic ecosystem mostly through wastewater discharges. The primary sources of wastewater discharge are municipal wastewater treatment plants and industrial effluents. Municipal wastewater (domestic wastewater or sewage) treatment plants are a major source of developing pollutants (Egbuna et al., 2021). Pollutants including POPs and heavy metals that are not eliminated throughout the treatment operations are frequently present in treated wastewater. Domestic sewage and, to a lesser extent, industrial wastewater are treated at these facilities (Egbuna et al., 2021). Regardless of treatment methods, not all EPs are adequately eliminated. Some may pass through treatment or resist deterioration. Chemicals, pharmaceuticals, and microplastics, among other new contaminants, can be released into water bodies by industrial operations. These contaminants are frequently dumped into neighboring rivers, lakes, or seas because most people see the aquatic bodies as a repository of wastes (Geissen et al., 2015).

4.4 EMERGING POLLUTANTS IMPACT PLANKTON AT THE INDIVIDUAL LEVEL

4.4.1 Physiological Disruptions

Some emerging pollutants possess endocrine-disrupting substances that can disrupt plankton reproduction and reduced reproduction rates, changing sex ratios, and interrupted life cycles can all result from this (Pico et al., 2019). A variety of marine organisms, including zooplankton, are impacted by EPs in terms of their fertility, metabolism, and mortality (Pico et al., 2019). When exposed to specific EPs, plankton's feeding habits may be modified. For example, microplastics may impact zooplankton by lowering feeding and energy reserves, depriving them of nutrients, and causing physical damage to body parts and the digestive system (He et al., 2021). Exposure to pollutants can impair their capacity to acquire and consume food, resulting in lower feeding efficiency. When exposed to EPs, plankton may become more toxic and exhibit stress reactions. Pollutants can cause oxidative stress and harm organelles, tissues, and cells in plankton, which can have an impact on their general well-being (Squadrone et al., 2021).

4.4.2 Behavioral Impacts

Behavioral responses such as frequency of movements and swimming speed and trajectory are particularly important to consider when utilizing methods that make use of the most dynamic markers of plankton community health (Dyomin et al., 2023). Planktonic organisms use particular cues to navigate and find their way around their immediate environment (Dyomin et al., 2023). Certain EPs, like microplastics and surfactants, can interfere with these cues, making it harder for the organisms to stay vertical in the water column and causing perplexity. Plankton's sensory and feeding organs may be impacted by EPs, which may hinder their capacity to effectively find and catch prey. This may affect their ability to grow and survive and cause nutritional stress.

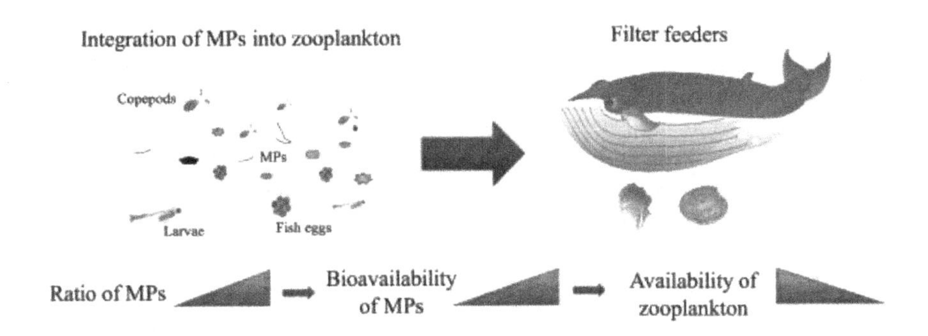

FIGURE 4.2 The incorporation of microplastics into zooplankton diminishes its capacity to filter feeders.

Source: He et al. (2021).

4.4.3 Bioaccumulation

The ability of EPs to persist in aquatic environments causes bioaccumulation and potential harm to the plankton community, and there is a chance that some EPs will bioaccumulate and biomagnify in plankton. This implies that over time, they may build up in the tissues of these organisms to the point where concentrations may be hazardous to the plankton itself or to other organisms that eat it, like small fish and other zooplankton (Wang et al., 2017). Through direct uptake or by eating contaminated food particles, plankton can accumulate EPs, which can alter the dynamics and structure of entire aquatic communities (Yang et al., 2020). The accumulation of EPs in phytoplankton has unexpected consequences that impact the food web and ultimately impact the lives of higher trophic organisms. Red tides may occur as a result of an increase in phytoplankton biomass (Xin et al., 2020). Toxins produced by red tides can build up in higher trophic organisms like clams and oysters and make them dangerous to eat by humans. Severe and even fatal neurological consequences can result from toxins in oysters (Xin et al., 2020). The consumption of microplastics by these organisms has also been linked in studies to reduced growth, longevity, fertility, and reproduction in a range of plankton species (He et al., 2021). Also, when microplastics are assimilated into zooplankton, it reduces the ability of filter feeders to sieve their food properly before they ingest them (Figure 4.2). When MPs are consumed by zooplankton, either directly or indirectly, the result is altered or even deadly gene expression, intestinal injury, decreased intake, sluggish or delayed growth, and decreased spawning. Zooplankton's behavior, procreation, and progeny are likewise impacted by MP bioavailability.

4.5 EFFECTS OF EMERGING POLLUTANTS ON PLANKTON AT THE COMMUNITY LEVEL

4.5.1 Changes in Community Structure

Changes in the relative abundance of various plankton species within a community can result from exposure to EPs. Certain species might withstand pollution better than others, while others might suffer harm or become extinct (Drago and Weithoff, 2021). Though cladocerans are more susceptible to plastic contamination, freshwater zooplankton such as rotifers may be less susceptible (Drago and Weithoff, 2021). Algae are utilized as indicators of water contamination in various water bodies or to gauge the level of pollution. The majority of green algae are freshwater species, which might not be able to withstand pollution. According to Jafari and Gunale (2006), *Chlorophyceae, Bacillariophyceae, Cyanophyceae*, and *Euglenophyceae* were among the genera of phytoplankton that could withstand pollution. While some species of phytoplankton—*Nitzschia palea, Synedra acus, Scenedesmus quadricauda, Pandorina morum*, and *Trachelomonas volvocina*—were found to be tolerant of pollution (Jafari and Gunale, 2006).

The diversity of macroinvertebrates in the marine ecosystem is severely harmed by EPs, and it has been found that many of these pollutants specifically change the structure of the macroinvertebrate community (Sumudumali and Jayawardana, 2021). Macroinvertebrates are essential to the health of aquatic ecosystems because they transfer energy to higher trophic levels in an aquatic food web (Sumudumali and Jayawardana, 2021). Emerging pollutants have also been linked to the transboundary movement of microbial species, disrupting ecosystem function and community structure (Rosato et al., 2020). EPs may cause changes in population numbers and species diversity in plankton communities. Pollutant exposure in aquatic systems may result in altered population numbers (particularly extinctions) and, as a result, altered community composition and species richness, which are known to impact ecosystem functioning (Brodin et al., 2014). Pollutants may cause some species to thrive, leading to domination, while others may decrease or vanish. This has the potential to upset the community's balance and stability.

4.5.2 Impact on Trophic Interactions

EPs have the potential to alter species composition and abundance within the biota, leading to competition between more resilient species and the eventual extinction of the most vulnerable ones (Duarte et al., 2023). Changes in the organization of the plankton population can disturb the complicated food web linkages in aquatic environments. For example, a decrease in herbivorous zooplankton abundance caused by EP exposure may cause a decrease in the amount of herbivorous zooplankton, which may then raise phytoplankton biomass and possibly cause algal blooms. Exposure to emerging contaminants may cause growth inhibition, reduced biomass, deformed cells, reduced cell sizes, and inhibition of photosynthesis (Chia et al., 2021). For many aquatic creatures, plankton is their main food source. Changes in the makeup of the plankton community may have an impact on these higher trophic levels, affecting fish populations and the general health of the ecosystem (Chia et al., 2021).

4.5.3 Promotion of Harmful Algal Blooms

The rapid growth of harmful algae on water surfaces is known as a harmful algal bloom (HAB) (Amorim and Moura, 2021). Harmful algal blooms may result from EPs' promotion of the growth of dangerous algal species (Amorim and Moura,

2021). Certain EPs can serve as fertilizers for harmful algae, such as runoff from agricultural areas that are rich in nutrients. These blooms can outcompete other plankton species, lower oxygen concentrations, and produce toxins that are hazardous to human health as well as aquatic life. Because algal blooms are inseparable from phytoplankton and very sensitive to phytoplankton dynamics, they produce imbalances in water quality and aquatic community structure, primarily impacting zooplankton biodiversity and community structure (Tian et al., 2023).

4.5.4 Disruption of Nutrient Cycling

The nutrient cycle in aquatic ecosystems may be disrupted by changes in plankton communities impacted by EPs. Water quality and ecosystem productivity may be impacted, for instance, by changes in the dominance of particular plankton species, which may have an impact on the uptake and recycling of nutrients like phosphorus and nitrogen (Baho et al., 2019).

4.6 ELIMINATION STRATEGY OF EPS IN THE ENVIRONMENT

Researchers are becoming more and more concerned about the treatment and removal of EPs from the environment due to the ongoing increase in awareness of their presence and the threats they pose (Egbuna et al., 2021). Numerous treatment techniques, such as physico-chemical, oxidation, and biological techniques, can be used to eliminate EPs, and most treatment technologies have drawbacks, such as secondary pollution, expensive maintenance costs, and difficult treatment procedures. Conventional wastewater treatment techniques remove soluble and insoluble contaminants through sewage removal operations and physical, chemical, and biological processes (Crini and Lichtfouse, 2019). It is challenging to remove EPs from industrial and municipal wastewater treatment plants using traditional treatment methods because of their diverse constituents. Adsorption, membrane technology, accelerated oxidation process, biological approaches, and wetlands (natural engineered wastewater treatment systems) are also some of the technologies used to remove EPs to create treatment solutions (Vieira et al., 2020).

The main components of the physicochemical process are membrane filtration, coagulation-flocculation, and activated carbon. Adsorption with activated carbon is used to remediate industrial wastewater (Mishra et al., 2023). The adsorption process is popular and significant in the water treatment industry because it is simple to design, produces no undesirable by-products, and is resistant to dangerous substances (Mishra et al., 2023). When employing chemical oxidants like ozone or chlorine to treat emerging pollutants, oxidation is an essential technique. By-products can form as a result of extremely reactive reactions in water. Because of this, when choosing this treatment, chemical oxidants must be carefully chosen (Wang et al., 2023). During decomposition, O_3 either directly or indirectly reacts with organic pollutants to produce molecular O_3 and free radicals. For breaking down and getting rid of pollutants, the most economical biological technique is conventional activated sludge (CAS). While most of the components in this method are soluble in the effluent, some are absorbed in the sludge. Membrane bioreactors (MBRs) represent an additional effective biological treatment option for bulk organics removal. These systems can function as hybrids or in tandem with conventional CAS (Arman et al., 2021). The capacity of MBRs to treat different wastewater compositions is one of their main advantages over CAS (Arman et al., 2021).

Currently, EPs are removed from water using adsorption, oxidation, or a mix of biological and technological treatment procedures. Because of its versatility, ease of use, and disregard for hazardous chemicals, the adsorption approach is one of the most effective and reliable methods for treating wastewater (Rathi et al., 2021). Adsorption is advantageous because it does not produce undesirable by-products and allows for appropriate treading of the loaded adsorbed material. Advanced oxidation processes are unique approaches to the elimination of EPs that have several advantages over previous methods. Other processes like sonolysis, Fenton's reaction, ozonation, UV radiation, photocatalysis, microwave radiation, and electrolysis are combined in this process (Mishra et al., 2023). Additionally, constructed wetlands are an inexpensive wastewater treatment option that can help reduce EPs (Chen et al., 2016).

4.7 CONCLUSION

Emergent contaminants pose a momentous danger to plankton populations and the overall health of aquatic ecosystems. EPs may cause detrimental algal blooms, alter the structure of the community, and interfere with the nutrient cycle. The environmental effects of emerging contaminants on plankton populations provide a thorough and enlightening assessment of the developing risks to plankton populations posed by contaminants. Knowing the ecological effects of these pollutants and implementing efficient management strategies are essential to maintaining the biodiversity and healthy functioning of the planktonic community. The critical importance of plankton in aquatic ecosystems and the urgency of taking urgent action to maintain their fragile existence are necessary. Therefore, stringent measures (e.g. standard erosion and sediment control measures, use of dishwasher and detergent without phosphates, proper disposal of toxic chemicals and medical waste) should be implemented on the discharge of EPs to the aquatic habitat to safeguard the plankton population.

REFERENCES

Amorim, C.A., and Moura, A.D.N. (2021). Ecological impacts of freshwater algal blooms on water quality, plankton biodiversity, structure, and ecosystem functioning. *Science of the Total Environment*, 758, 143605.

Arienzo, M., Ferrara, L., and Trifuoggi, M. (2021). The dual role of microplastics in marine environment: Sink and vectors of pollutants. *Journal of Marine Science & Engineering*, 9(6), 642.

Arman, N.Z., Salmiati, S., Aris, A., Salim, M.R., Nazifa, T.H., Muhamad, M.S., and Marpongahtun, M. (2021). A review on emerging pollutants in the water environment: Existences, health effects and treatment processes. *Water*, 13(22), 3258.

Baho, D.L., Pomati, F., Leu, E., Hessen, D.O., Moe, S.J., Norberg, J., and Nizzetto, L. (2019). A single pulse of diffuse contaminants alters the size distribution of natural phytoplankton communities. *Science of the Total Environment*, 683, 578–588

Batten, S.D., Abu-Alhaija, R., Chiba, S., Edwards, M., Graham, G., Jyothibabu, R., and Wilson, W. (2019). A global plankton diversity monitoring program. *Frontiers in Marine Science*, 6, 321. https://doi.org/10.3389/fmars.2019.00321

Brodin, T., Piovano, S., Fick, J., Klaminder, J., Heynen, M., and Jonsson, M. (2014). Ecological effects of pharmaceuticals in aquatic systems—impacts through behavioural alterations. *Philosophical Transactions of the Royal Society of London. Series B, Biological Sciences*, 369(1656), 20130580.

Calvo-Flores, F.G., Isac-Garcéa, J., and Dobado, J.A. (2017). Industrial chemicals as emerging pollutant. In *Emerging Pollutants*, Wiley-VCH, pp. 265–340. https://doi.org/10.1002/9783527691203.ch9

Cavan, E.L., Henson, S.A., Belcher, A., and Sanders, R. (2017). Role of zooplankton in determining the efficiency of the biological carbon pump. *Biogeosciences*, 14(1), 177–186.

Chen, Y., Vymazal, J., Březinová, T., Koželuh, M., Kule, L., Huang, J., and Chen, Z. (2016). Occurrence, removal and environmental risk assessment of pharmaceuticals and personal care products in rural wastewater treatment wetlands. *Science of the Total Environment*, 566–567, 1660–1669.

Chia, M.A., Lorenzi, A.S., Ameh, I., Dauda, S., Cordeiro-Araújo, M.K., Agee, T., Okpanachi, I.Y., and Adesalu, T. (2021). Susceptibility of phytoplankton to the increasing presence of active pharmaceutical ingredients (APIs) in the aquatic environment: A review. *Aquatic Toxicology*, 234, 105809.

Crini, G., and Lichtfouse, E. (2019). Advantages and disadvantages of techniques used for wastewater treatment. *Environmental Chemistry Letters*, 17, 145–155.

Drago, C., and Weithoff, G. (2021). Variable fitness response of two rotifer species exposed to microplastics particles: The role of food quantity and quality. *Toxics*, 9(11), 305.

Duarte, J.A.P., Ribeiro, A.K.N., de Carvalho, P., Bortolini, J.C., and Ostroski, I.C. (2023). Emerging contaminants in the aquatic environment: Phytoplankton structure in the presence of sulfamethoxazole and diclofenac. *Environmental Science & Pollution Research*, 30, 46604–46617.

Dyomin, V., Morgalev, Y., Polovtsev, I., Morgalev, S., Morgaleva, T., Davydova, A., and Kondratova, O. (2023). Phototropic behavioral responses of zooplankton in Lake Baikal in situ and during the anthropogenic impact modeling. *Water*, 15(16), 2957. https://doi.org/10.3390/w15162957

Ebenezer, V., Nancharaiah, Y.V., and Venugopalan, V.P. (2012). Chlorination-induced cellular damage and recovery in marine microalga, Chlorella salina. *Chemosphere*, 89(9), 1042–1047.

Egbuna, C., Amadi, C.N., Patrick-Iwuanyanwu, K.C., Ezzat, S.M., Awuchi, C.G., Ugonwa, P.O., and Orisakwe, O.E. (2021). Emerging pollutants in Nigeria: A systematic review. *Environmental Toxicology & Pharmacology*, 85, 103638.

El-Bassat, R.A., Touliabah, H.E., Harisa, G.I., and Sayegh, F.A.Q. (2011). Aquatic toxicity of various pharmaceuticals on some isolated plankton species. *International Journal of Medical Science*, 3(6), 170–180.

Emegha, J.O., Oliomogbe, T.I., Okpoghono, J., Babalola, A.V., Ejelonu, C.A., Elete, D.E., and Ukhurebor, K.E. (2023). Green biosorbents for the degradation of petroleum contaminants, chapter 12. In *Adsorption Applications for Environmental Sustainability*, IOP Science, pp. 12–20. https://doi.org/10.1088/978-0-7503-5598.

Geissen, V., Mol, H., Klumpp, E., Umlauf, G., Nadal, M., van der Ploeg, M., van de Zee Sjoerd, E.A.T.M., and Ritsema, C.J. (2015). Emerging pollutants in the environment: A challenge for water resource management. *International Soil & Water Conservation Research*, 3(1), 57–65.

He, M., Yan, M., Chen, X., Wang, X., Gong, H., Wang, W., and Wang, J. (2021). Bioavailability and toxicity of microplastics to zooplankton. *Gondwana Research*. https://doi:10.1016/j.gr.2021.07.021

Jafari, N.G., & Gunale, V.R. (2006). Hydrobiological study of algae of an urban freshwater river. *Journal of Applied Sciences & Environmental Management*, 10(2), 153–158.

Krishnakumar, S., Singh, D.S.H., Godson, P.S., and Thanga, S.G. (2022). Emerging pollutants: Impact on environment, management, and challenges. *Environmental Science & Pollution Research*, 29, 72309–72311.

Li, J., Wang, Y., Li, N., He, Y., Xiao, H., Fang, D., and Chen, C. (2022). Toxic effects of bisphenol A and bisphenol S on chlorella pyrenoidosa under single and combined action. *International Journal of Environmental Research & Public Health*, 19(7), 4245.

Liu, S., Wang, C., Wang, P., Chen, J., Wang, X., and Yuan, Q. (2021). Anthropogenic disturbances on distribution and sources of pharmaceuticals and personal care products throughout the Jinsha River Basin, China. *Environmental Research*, 198, 110449.

Lobus, N.V., and Kulikovskiy, M.S. (2023). The co-evolution aspects of the biogeochemical role of phytoplankton in aquatic ecosystems: A review. *Biology*, 12(1), 92. https://doi.org/10.3390/biology12010092

Lomartire, S., Marques, J.C., and Gonçalves, A.M.M. (2021). The key role of zooplankton in ecosystem services: A perspective of interaction between zooplankton and fish recruitment. *Ecological Indicators*, 129, 107867.

Mahmood, I., Imadi, S.R., Shazadi, K., Gul, A., and Hakeem, K.R. (2016). Effects of pesticides on environment. In Hakeem, K., Akhtar, M., and Abdullah, S., editors. *Plant, Soil and Microbes: Volume 1: Implications in Crop Science.* Cham: Springer International Publishing, pp. 253–269.

Mishra, R.K., Mentha, S.S., Misra, Y., and Dwivedi, N. (2023). Emerging pollutants of severe environmental concern in water and wastewater: A comprehensive review on current developments and future research. *Water-Energy Nexus*, 6, 74–95.

Mojiri, A., Nazari, V.M., Ansari, H.K., Vakili, M., Farraji, H., and Kasmuri, N. (2023). Toxicity Effects of Perfluorooctanoic Acid (PFOA) and PerfluorooctaneSulfonate (PFOS) on two green microalgae species. *International Journal of Molecular Sciences*, 24(3), 2446.

Narayanan, M., El-sheekh, M., Ma, Y., Pugazhendhi, A., Natarajan, D., Kandasamy, G., Raja, R., Kumar, S.R.M., Kumarasamy, S., Sathiyan, G., Geetha, R., Paulraj, B., Liu, G., and Kandasamy, S. (2022). Current status of microbes involved in the degradation of pharmaceutical and personal care products (PPCPs) pollutants in the aquatic ecosystem. *Environmental Pollution*, 300, 118922. https://doi.10.1016/j.envpol.2022.118922

Okpoghono, J., George, B.O., Achuba, F.I., Igue, U.B., Okom, S.U., Seigha, A.A., Ugwuka, L.C., and Ayogoi, K.I. (2021). Impact of crude petroleum oil and monodora myristica on membrane bound ATPases and erythrocyte osmotic fragility in-vivo. *Uniport Journal of Engineering & Scientific Research*, 6(1), 65–77.

Osier, M.N., and Cohen, R.A. (2020). Mixtures of tetrakishydroxymethylphosphonium chloride and ammonium influence plankton community structure in blackwater pond mesocosms. *Emerging Contaminants*, 6, 62–71.

Osuoha, J.O., Anyanwu, B.O., and Ejileugha, C. (2023). Pharmaceuticals and personal care products as emerging contaminants: Need for combined treatment strategy. *Journal of Hazardous Materials*, 9, 10206.

Pico, Y., Alfarhan, A., and Barcelo, D. (2019). Nano- and microplastic analysis: Focus on their occurrence in freshwater ecosystems and remediation technologies. *Trends in Analytical Chemistry*, 113, 409–425.

Pullagurala, V.L.R., Rawat, S., Adisa, I.O., Hernandez-Viezcas, J.A., Peralta-Videa, J.R., and Gardea-Torresdey, J.L. (2018). Plant uptake and translocation of contaminants of emerging concern in soil. *Science of the Total Environment*, 636, 1585–1596.

Rajmohan, K.S., Chandrasekaran, R., and Varjani, S. (2020). A review on occurrence of Pesticides in environment and current technologies for their remediation and management. *Indian Journal of Microbiology*, 60(2), 125–138.

Rathi, B.S., Kumar, P.S., and Show, P.L. (2021). A review on effective removal of emerging contaminants from aquatic systems: Current trends and scope for further research. *Journal of Hazardous Materials*, 409, 124413.

Rosato, A., Barone, M., Negroni, A., Brigidi, P., Fava, F., Xu, P., Candela, M., and Zanaroli, G. (2020). Microbial colonization of different microplastic types and biotransformation of sorbed PCBs by a marine anaerobic bacterial community. *Science of the Total Environment*, 705, 135790.

Roy, T., and Martin, V. (2014). Medium-chain chlorinated paraffins (MCCPs): A review of bioaccumulation potential in the aquatic environment. *Integrated Environmental Assessment & Management*, 10(1), 78–86.

Saquib, S., Yadav, A.K., and Prajapat, K.B. (2021). Emerging pollutants in water and human health. In Ahamad, A., Siddiqui, S.H., and Singh, P., editors. *Contamination of Water: Health Risk Assessment and Treatment Strategies*. Academic Press, pp. 285–299.

Squadrone, S., Pederiva, S., Bezzo, T., Sartor, R.M., Battuello, M., Nurra, N., Griglione, A., Brizio, P., and Abete, M.C. (2021). Microplastics as vectors of metals contamination in Mediterranean Sea. *Environmental Science and Pollution Research*, 29(20), 29529–29534.

Sumudumali, R.G.I., and Jayawardana, J.M.C.K. (2021). A review of biological monitoring of aquatic ecosystems approaches: With special reference to macroinvertebrates and pesticide pollution. *Environmental Management*, 67, 263–276.

Tian, X., Yuan, Y., Zou, Y., Qin, L., Zhu, X., Zhu, Y., Zhao, Y., Jiang, M., and Jiang, M. (2023). Cyanobacterial blooms increase functional diversity of metazooplankton in a shallow eutrophic lake. *Water*, 15(5), 953.

Vieira, W.T., de Farias, M.B., Spaolonzi, M.P., da Silva, M.G.C., and Vieira, M.G.A. (2020). Vieira Removal of endocrine disruptors in waters by adsorption, membrane filtration and biodegradation. A review. *Environmental Chemistry Letters*, 18, 1113–1143.

Wang, H., Wang, Y., and Dionysiou, D.D. (2023). Advanced oxidation processes for removal of emerging contaminants in water. *Water*, 15(3), 398.

Wang, Q., Chen, M., Shan, G., Chen, P., Cui, S., Yi, S., and Zhu, L. (2017). Bioaccumulation and biomagnification of emerging bisphenol analogues in aquatic organisms from Taihu Lake, China. *Science of the Total Environment*, 598, 814–820.

Wu, P., Huang, J., Zheng, Y., Yang, Y., Zhang, Y., He, F., Chen, H., Quan, G., Yan, J., Li, T., and Gao, B. (2019). Environmental occurrences, fate, and impacts of microplastics. *Ecotoxicology & Environmental Safety*, 184, 109612.

Xin, X., Huang, G., and Zhang, B. (2020). Review of aquatic toxicity of pharmaceuticals and personal care products to algae. *Journal of Hazardous Materials*, 410(7), 124619.

Yang, H., Lu, G., Yan, Z., Liu, J., Dong, H., Bao, X., Zhang, X., and Sun, Y. (2020). Residues, bioaccumulation, and trophic transfer of pharmaceuticals and personal care products in highly urbanized rivers affected by water diversion. *Journal of Hazardous Materials*, 391, 122245.

5 Occurrence and Fate of Pharmaceutical and Cosmetic Wastes on Plankton Consortia

Ebere Mary Eze, Precious Onome Obiebi, Joel Okpoghono, Oke Aruorem, Peter Mudiaga Etaware, Odiri Ukolobi, Uduenevwo Francis Evuen, and Joshua Othuke Orogu

5.1 INTRODUCTION

Organisms called planktons (singular plankter) are unable to swim against currents and are found in the water column. From tiny to macroscopic species, planktons span a broad spectrum of sizes, such as jellyfish, algae, etc. (Zhang et al., 2018). Plankton is made up of prokaryotic or eukaryotic cell structures seen in unicellular, filamentous, or colonial organisms (Smith, 2013). These creatures are extremely important to the ecology of the various water bodies found throughout the planet because they repair carbon dioxide, produce oxygen, and have an essential component serving as the foundation for many food systems (Pierella et al., 2020). Plankton can be divided in two different ways. They can be divided according to the type of organisms: bacterioplankton, phytoplankton (predominantly autotrophic organisms), and zooplankton (predominantly heterotrophic organisms) (Pierella et al., 2020). The can also be divided by size: femtoplankton (0.02–0.2 μm), picoplankton (0.2–2.0 μm), nanoplankton (2.0–20 μm), microplankton (20–200 μm), mesoplankton (0.2–20 mm), macroplankton (2–20 cm), and megaplankton (20–200 cm) (Pierella et al., 2020). Planktonic organisms are found in all types of waters in which they have suitable conditions for survival and food source (e.g. on glaciers, puddles, rivers, in the ocean).

Pharmaceuticals, a milestone in medical science, have saved lives, cured millions of diseases, and improved the quality of life. This success has now led to continuous aquatic pollution, which could endanger both the environment and human health (Ngqwala and Muchesa, 2020). Pharmaceuticals and cosmetics are broad classes of compounds that usually cause uncommon pollutants in aquatic environments. Although pharmaceuticals and cosmetics exist at low concentrations in aquatic environments. They have the following characteristics in the aquatic ecosystem:

DOI: 10.1201/9781003362975-5

pseudo-persistence and bioaccumulation which happens as a consequence of low degradation ability, extensive use, and continuous release (Ngqwala and Muchesa, 2020). Medications used in humans that include hormones, antibiotics, analgesics, antidepressants, and antineoplastics and veterinary medications that include hormones, antibiotics, and parasiticides are the main pharmaceutical products that are identified as posing an environmental risk (Küster and Adler, 2014).

Late in the thirties, natural and synthetic antibiotics were introduced, and their usage has increased for human and animal production. They are regarded to be pseudo-persistent, meaning that they enter the aquatic ecosystem continuously and remain permanently present. The different classes of antibiotics used in medical sciences are ß-lactams, macrolides, fluoroquinolones, aminoglycosides, sulfonamide, and tetracycline (Smith and Johnson, 2023). The most used antibiotics are the ß-lactams which consist of amoxicillin and penicillin. These are used for human therapy (Smith and Johnson, 2023). In veterinary medicine, antibiotics are used to treat disease or to increase feed efficiency and improve growth rates, for example, in shrimp hatcheries and cultural ponds (Martel et al., 2021). Antihypertensive are the drugs used to treat high blood pressure. There are many kinds of antihypertensive, for instance, ß-blockers, calcium channel blockers, and diuretics. The most common central nervous system and metabolic stimulant is caffeine. It is mostly present in many drinks and some pharmaceuticals to reduce physical fatigue and to regain alertness when drowsiness occurs. Also, caffeine is easily found in surface water (Kolpin et al., 2002).

Due to their widespread usage and detrimental effects on both humans and other animals, hormonal chemicals are one of the most significant families of pharmaceuticals. The most important natural estrogen includes estriol, estradiol, and estrone which are mainly excreted from human beings. Another important synthetic estrogen such as 17a-ethinylestradiol which is used for contraception by women causes critical effects on the environment such as feminization of male fish, alternation of DNA integrity, immune cell number, and ability to break pollutants (Kolpin et al., 2002).

5.2 PHARMACEUTICAL WASTE: TYPES AND SOURCES OF POLLUTION

Pharmaceutical waste can be defined as an expired or unused/leftover medication that can be classified as non-hazardous or hazardous, depending on its chemical properties. According to Weston and Caminada (2006), pharmaceutical wastes comprise waste products from manufacturers' processes as well as over-the-counter pharmaceuticals and prescription drugs. Moreover, anything that comes into touch with medications, such as sharp objects, and personal protective equipment (PPE), like masks and gloves, might be considered as pharmaceutical waste.

5.2.1 Sources of Pharmaceutical Waste

Production and distribution of cosmetics and pharmaceuticals are primarily driven by the growing requirements of the human population discharge into the environmental matrices (water, soil sediment, and wastewater) (Hai Nguyen and Reinhard, 2010). Some pharmaceutical compounds are consumed, metabolized, and excreted into domestic sewage as a mixture of parent compounds and their metabolites which seep through septic tanks, open drainages, drainage pipes, and runoff surface water into water bodies made up of conjugated glucuronides and transformation products (Heberer, 2002).

Preliminary research has demonstrated that conjugated pharmaceutical compounds readily cleave during wastewater treatment, releasing the component into the treated wastewater and eventually into the environment (Ternes, 1998). Pharmaceutical compounds and cosmetic wastes are distributed into the environment by two major means: the point sources and the diffuse sources. The diffused sources are pollutants discharged from large areas and combined into several points and hardly recognized as distinct sites. Diffuse pollution is preferable to point sources of pollution since it often has less environmental loading because it is naturally attenuated in shallow and subsurface soil. On the other hand, point sources of pollution are the separate sites releasing the pollutants from a particular point and the number of pollutants released through them can be measured easily via mathematical modeling. For instance, treated sewage from industries, hospitals, and septic tanks is the point source for the soil zone and the water resources. The point sources are further grouped into four, namely, pharmaceutical manufacturing plants, healthcare institutions, external care facilities, personal care products, and veterinary offices.

5.3 PHARMACEUTICAL MANUFACTURING PLANTS

One common source of pharmaceutical waste is pharmaceutical manufacturing facilities. In pharmaceutical production facilities, physical wastes are among the wastes produced. It could be leftover goods, empty chemical containers containing residues, or the cloths and mop heads used to clean tools and clean up spills (Gurudatta et al., 2020). Some pharmaceutical companies dispose of this waste in a landfill, where the chemicals may leak into the groundwater and soil and endanger the ecosystem or contaminate the nearby water supply (Geiger et al., 2016). Pharmaceutical wastes when emptied into the drain can negatively affect the treatment of surface water, contaminate local bodies of water, and harm wildlife. To prevent endangering public health, damaging the environment, and facing harsh fines, the plant must dispose of this trash by tight guidelines. Certain chemical residues are left on equipment surfaces in drug manufacturing factories during the drug-processing process. Pharmaceutical residues pollute the water used in manufacturing plants' cleaning procedures when they drain and clean holding tanks or clean their contaminated surfaces (Germain et al., 2018). To protect human health and the environment, facilities must also appropriately dispose of the wastewater created by the polluted water.

5.3.1 Healthcare Institutions and Extended Care Facilities

Healthcare and extended care facilities use numerous pharmaceutical products every day. Certain products, like discarded syringes, empty prescription bottles, expired medications, and other items tainted with pharmaceutical substances, eventually end up in the trash (Goudard and Loreau, 2008). Many of these materials, if incorrectly disposed of by the institution, could pollute the environment or have negative health repercussions.

5.3.2 Personal Care Products

Manufacturers of personal care products like cosmetics, perfumes, creams, and lotions generate wastes that require specialized disposal techniques. Care must be taken when discarding empty chemical containers and unsold goods. Specialized wastewater disposal is required because the chemical residues left on production equipment after cleaning can contaminate runoff water (Goudard and Loreau, 2008).

5.3.3 Veterinary Offices

Veterinary offices use tremendous quantities of pharmaceutical products. An average veterinary office supplies a variety of a procedure higher than a human physician. So a veterinary clinic may produce a more extensive range of pharmaceutical waste higher than that of a human physician (Goudard and Loreau, 2008).

5.3.4 The Regulatory Bodies of Pharmaceutical Wastes

The main governing body that regulates the discarding of pharmaceutical waste is the Environmental Protection Agency (EPA). Additional regulatory bodies include the following: Drug Enforcement Agency (DEA), Department of Transportation (DOT), Joint Commission (JC), Occupational Health and Safety Administration (OSHA), and Fish and Wildlife Services (FWS).

5.4 THE CATEGORIES OF PHARMACEUTICAL WASTES

There are two primary classifications of pharmaceutical waste: solid waste and liquid waste.

5.4.1 Solid Pharmaceutical Waste

Used items containing pharmaceutical wastes are considered solid pharmaceutical waste. Sharp objects include syringes, scalpels, and needles. Items that are contaminated include bandages, gloves, masks, IV bags, and tubing. Medications comprise either non-hazardous or dangerous substances. Empty containers include ointment tubes, blister packs, pill bottles, and liquid medication containers. Medication delivery systems such as nebulizers, autoinjectors, and inhalers (Halling-Sbrensen et al., 1998).

5.4.2 Liquid Pharmaceutical Waste

Pharmaceutical manufacturing facilities generate liquid pharmaceutical wastes when they carry out specific processing procedures. Sludge from chemical processing, polluted solvents from tank cleaning, and leftover liquid medications are a few examples of these wastes (Halling-Sbrensen et al., 1998).

5.5 COSMETIC WASTES

Cosmetics are substances used by the human body that are intended to promote beauty, cleanse, beautify, or change appearance without compromising the body's structure or functions, according to the Food and Drug Administration (FDA) (Kolpin et al., 2002). Perfumes, moisturizers, nail polish, makeup, hair products, and even toothpaste fall under this category. In the cosmetics industry, packaging tends to be more whimsical than functional. Packaging seems to be a major focus for beauty firms. It's common to find more unnecessary packaging for products that are more costly or opulent. Packaging, which includes plastic, paper, glass, and metal, accounts for around 70% of the trash produced by the beauty industry. These materials wind up in landfills as well as rivers, lakes, and oceans (Kolpin et al., 2002). Plastic packaging is only one aspect of the plastic pollution caused by beauty products. Microbeads are plastic particles found in the items themselves (Kolpin et al., 2002). Body scrubs and other cosmetics contain microbeads as an exfoliator. They are made up of small plastic materials, not bigger than 5 mm. When a product containing microbeads is used, countless microplastics are sent down the drain into water bodies. There, they draw poisons, which fish and eventually people ingest. Microbeads exacerbate the severe plastic pollution that already plagues the world's oceans. Despite being outlawed in the US and the UK in 2015 and 2018, respectively, microbeads are still found in certain goods. Nothing is stopping cosmetic producers operating in nations without laws regulating microbeads from employing dangerous polymers. Parabens and triclosan are chemicals found in a variety of cosmetic products acting as preservatives. For example, some sunscreen products contain benzophenone or its derivative oxybenzone. These chemicals have been tagged as endocrine disruptors and linked with cancer (Gurudatta et al., 2020). Fourteen thousand tons of sunscreen get into the ocean and settle on coral reefs every year (Gurudatta et al., 2020). Oxybenzone is a culprit in the massive destruction of coral reefs in the world's oceans. Chemical ingredients like BHT, sodium laureth sulfate, and BHA have been found to cause changes to the biochemistry of aquatic life (Kolpin et al., 2002).

5.6 THE EFFECTS OF COSMETICS AND PHARMACEUTICAL WASTES ON PLANKTON COMMUNITIES

Drugs remain active after being discharged into the environment, so they can affect any aquatic organisms by interfering with their biological systems as enzymes; the effect varies according to the chemical structure of the drug. Lipophilic compounds may accumulate in sediments or soils while the mobility of water-soluble compounds may lead to contamination of groundwater (Isidori et al., 2007; Fent et al., 2006). Because of their physicochemical and biological properties, there are concerns about their potential impact on non-target species (Isidori et al., 2007). Data from the literature on the mechanisms governing the fate and effects of these active substances or their derivatives arising from drug transformations in the environment, which may be more hazardous than the original drug. DellaGreca et al. (2003) rarely provide qualitative and quantitative information. Pharmaceuticals have been detected in surface waters and occasionally in groundwater (Zuccato et al., 2006).

Empirical studies have demonstrated that medicinal compounds at environmental concentrations have the potential to impact aquatic life (Isidori et al., 2005). The disturbance they cause to the microbial life in surface waters was reported by Kümmerer (2001), while their effects at low concentrations on other organisms were examined by Halling-Sbrensen (2000). Pharmaceuticals enter the aquatic environment via excrements in their original form or metabolites originating from households and hospitals via sewage treatment plants (STP). They are often resistant to biodegradation since metabolic stability is necessary for pharmacological effect. Some of these or their metabolites are also highly water-soluble, and therefore, the removal in wastewater treatments for such compounds is limited.

Veterinary and human pharmaceuticals also enter the aquatic environment by the spreading of manure or sewage sludge on fields for agricultural use. In this way, the pharmaceuticals enter the groundwater. The plankton communities are indicators of pharmaceutical presence in aquatic systems. High levels of cell malformations and mortality rates at high pharmaceutical concentrations are suggested by the majority of studies on plankton and individual medications.

Furthermore, a lot of research on plankton highlights how drugs affect the physiology of the organism. For example, periphyton assemblages exposed to fluoxetine showed decreased assemblage respiration and primary production in response to a drop in chlorophyll concentration. Likewise, exposure to trimethoprim and tylosin decreases *Navicula pelliculosa's* ability to use light efficiently during photosynthesis. A decrease in cell density and photosynthesis was seen in model plankton (*Phaeodactylum tricornutum*) following exposure to fluoxetine. Both freshwater and marine ecosystems are affected by the decrease in cell biomass because it can lead to a decrease in primary oxygen, a reduction in food sources, and a decrease in the availability of essential fatty acids for higher trophic-level organisms.

Studies examining the effects of drugs on complete plankton assemblages and the higher ecosystems they sustain are lacking. Furthermore, the complexity of the natural aquatic environment is not well replicated in the majority of ecotoxicity investigations. For instance, the availability of nutrients, such as phosphorus and total nitrogen, and the intensity of the light have a major impact on plankton growth. The way that plankton reacts to pesticide treatment (Debenest et al., 2010) suggests that they may react differently to the same chemical exposure under various environmental variables (such as light and nutrients) (Xin et al., 2020). Higher plankton formation rates, higher protein contents, and higher rates of enzyme synthesis are all indicators of an adequate nutrition supply.

The increase in enzyme synthesis is crucial for ecotoxicological studies because some enzymes may be inactivated by pharmaceutical compounds while others may help to metabolize the compounds (Xin et al., 2020). The environmental toxicity of pharmaceuticals to plankton and their transport and fate in the aquatic ecosystem is also influenced by the presence of natural organic matter, temperature, and pH. Increased light, specifically ultraviolet radiation (UV-A and UV-C), and temperature can damage plankton's photosynthetic system, affect cell integrity, cause DNA damage, influence their ability to fix nitrogen, cause oxidative stress, impact their settle ability, and result in toxin synthesis (Xin et al., 2020). It has also been established that during light exposure (sunlight and UV radiation), plankton are more sensitive to exposure to specific compounds (e.g. atrazine) (Debenest et al., 2010). Natural organic matter can affect diatom assemblages, but it can also change the ecological effects of pharmaceuticals and affect how they behave in the environment and disperse through physicochemical interactions like charge transfer, covalent bonding, hydrogen bonding, hydrophobic adsorption, and ion exchange (Xin et al., 2020).

5.7 WASTE PRODUCTS FROM COSMETICS AND PHARMACEUTICALS IN AQUATIC ENVIRONMENTS

Emerging pollutants, often referred to as pharmaceutical and cosmetic wastes (PnCWs), pose a global danger and have been found in aquatic ecosystems, including drinking water, sediments, seawater, groundwater, and surface water (Hai Nguyen and Reinhard, 2010). Large quantities of them are manufactured each year as a result of their widespread use in the home and healthcare industries. The majority of antibiotics are consumed and eliminated as metabolic products in the form of urine and feces, while cosmetics wash out of various bathing areas. These cosmetic and pharmaceutical wastes wind up in sewage systems and eventually enter the aquatic environment through wastewater treatment plant (WWTP) effluent discharge, sewage leaks, or the dumping of unwanted or by getting rid of leftover or incomplete prescription drugs. Using fertilizers made from industrial waste and animal dung can contaminate soil, which could then cause the contaminants to leak into aquatic environments. The primary sources of drinking water subterranean water and surface water from rivers, dams, and streams may become contaminated as a result (Lagesson et al., 2016). This calls attention to the state of the drinking water in our surroundings.

Chronic impacts, such as changes in the metabolic and reproductive systems of non-targeted animals, might arise from persistent and continuous exposure to pharmaceutical and cosmetic residues in the aquatic environment (Cooper et al., 2008). Since infections caused by bacteria resistant to antibiotics are difficult to treat and some antibiotics remain in the environment for months at a time, antibiotics in the environment have the potential to create antibiotic-resistant microorganisms even at low concentrations, which poses a health risk to both humans and animals (Cooper et al., 2008). For plants and animals to survive, water is essential (Kidd et al., 2007). Thus, preserving the aquatic environment is essential to living a good and healthy life.

5.8 THE EFFECTS OF COSMETICS AND PHARMACEUTICAL WASTES ON SURFACE WATER ORGANISMS

It is impossible to ignore the detrimental effects of pharmaceutical and cosmetic waste on organisms found in surface water. Several studies have shown that subjecting bacteria in our water to antibiotics may cause them to become more resistant to such effects (Cooper et al., 2008).

The majority of medications produced are water soluble and resistant to biomagnification because they are intended to have biological effects in modest doses (Sherer, 2006). That is, because the impacts of the natural food chain cannot reach the larger animals that consume more, the levels within them increase rapidly as other substances accumulate within their fat. The most evident concern raised by the results of (Laxminarayan et al., 2013) study is that the high concentrations of broad-spectrum antibiotics may contribute to the emergence of antibiotic resistance.

The widespread use of antibiotics in both human and animal medicine contributes to the major global threat that multidrug-resistant organisms pose to public health. Antibiotic-containing industrial effluent has the potential to favor environmental bacterial populations that are resistant (Guardabassi et al., 2021). According to Cleuvers (2008), a combination of nonsteroidal anti-inflammatory drug toxicity was significantly higher against Daphnia, even at concentrations where the individual treatments had no or very little effect. Surprisingly, at concentrations where no impacts on survival were seen, reproduction was reduced by 100%. This indicates that an acute test conducted at the same concentrations would completely miss this detrimental effect on the Daphnia population.

5.9 INDIRECT EFFECTS

Various effects of pharmaceuticals on the environment may be experienced by various animals. It has been discovered that the movement of chemicals throughout the food web chain may have negative consequences. Renal failure and visceral gout were linked between 2000 and 2003 to the significant yearly adult and sub-adult mortality (5%–86%) in the oriental white-backed vulture and the ensuing population reductions (34%–95%). Renal failure and the anti-inflammatory medication diclofenac residues were found to be directly correlated. By giving diclofenac-treated cattle to vultures and exposing them directly to the drug, researchers were able to replicate renal illness and diclofenac residues in oriental white-backed vultures (Oaks et al., 2010). According to other research, the dramatic losses in vulture populations are most likely the result of diclofenac use in veterinary medicine (Taggart et al., 2015).

Another example of indirect effects of antibiotics was reported by Hahn and Schulz (2007). The results of food selection studies using *Gammarus pulex* showed that leaves conditioned in the absence of two antibiotics, oxytetracycline and sulfadiazine, clearly preferred those conditioned in their presence. Other examples are the intersex of fish that is exposed to estrogen, and this feminization can severely disturb a fish population (Kidd et al., 2007). Pharmaceuticals in the aquatic system can also affect the behavior of fish, and this has been proved by Brodin et al. (2013), who found that oxazepam (a drug to treat anxiety) altered behavior and feeding rate among European perch even in concentration levels found in natural surface waters.

5.10 TOXIC EFFECTS OF PHARMACEUTICALS AND COSMETICS ON AQUATIC MICROORGANISMS

Sanyal et al. (2016) said that ibuprofen, which has analgesic, anti-inflammatory, and antipyretic properties (Reynolds, 2019) and is taken orally to treat mild to moderate pain of rheumatism and other musculoskeletal disorders, draws attention to the potential antimicrobial activity of ibuprofen against certain dermatophyte fungi. The same authors also noted that *Staphylococcus aureus* was susceptible to ibuprofen. Ibuprofen concentrations above 150 mg/L inhibited the development of *Staphylococcus aureus* at the starting pH of 7, but at pH 6, these amounts stopped the growth. Furthermore, Lee and Bird (2021) found that the *Calanoid copepods, Temoraturb inata,* if raised in pharmaceutical waste concentrations above 1 ppm, resulted in smaller adult size and diminished egg output as well as an irregular development pattern.

5.10.1 Phytoplankton

In a study conducted by Harrass et al. (2018), streptomycin inhibited the growth of six species of blue-green algae at doses ranging from 0.09 to 0.86 mg/1. While *Chlamydomonas reinhardtii* growth was inhibited at concentrations of 0.66 mg/L of active streptomycin, *Chlorella vulgaris*, *Scenedesmus obliquus*, and *Ulothrix sp.* grew in concentrations less than 21 mg/L. Sub-lethal dosages of streptomycin inhibited or delayed algal development, and some species' maximum densities were lowered.

5.10.2 Amphipods—Invertebrates

Lee and Bird (2021) examined the harmful consequences on the marine amphipod *Ampithoe valida* of pharmaceutical pollutants deposited into the water. Exposure to waste concentration for longer periods resulted in more harmful consequences. Comparing chronically exposed amphibian populations to waste concentrations exceeding 1%, the latter group exhibited decreased fertility and worse survival rates. In contrast to those exposed to less than 2% waste, which were able to survive for more than two months, the parent amphipods subjected to 3% garbage died 100% after three weeks. As a result, depending on the species used in the bioassay, the toxic levels were between 0.05% and 5% of the waste concentration. According to Nicol and Miller (2020), numerous invertebrates were poisonous to the pharmaceutical wastes disposed of at the Puerto Rico dumpsite; hence, depending on the species used in the bioassay, the toxic levels were between 0.05% and 5% of the waste concentration. Lagesson et al. (2016) found that five pharmaceutical substances were absorbed by four different aquatic invertebrate taxa (damselfly larvae, mayfly larvae, water louse, and ramshorn snails). They also found that there were notable variations in the ability of each drug to bioaccumulate, as well as in the uptake of different species.

5.10.3 Fish

Only little information is outlined in the literature concerning the effects of medical substances on fish species. Lagesson et al. (2016) conducted a recent study to determine the effects of certain pharmaceuticals. They measured the bioconcentrations of five pharmaceuticals—diphenhydramine, oxazepam, trimethoprim, diclofenac, and hydroxyzine—over several months in a large-scale, semi-natural pond system to determine the extent to which fish (European perch) ingest these substances. The findings imply that there are notable variations in both the ability of medications to bioaccumulate and in the absorption of different species.

5.10.4 Insects

Macri et al. (2018) showed that furazolidone had a significant toxic effect on mosquito larvae. However, formulations of coumaphos, dichlorvos, and phenothiazine adversely reduced their survival and reproduction for at least four to five days after treatment. In contrast, medications like piperazine, thiabendazole, and levamisole had little to no effect on clinging beetle breeding. Phenothiazine was also associated with deleterious changes in the botanical composition of pastures, whereas residues of dichlorvos delayed dung degradation (Lumaret, 2016). In the late eighties, Wall and Strong (2017) discovered that ivermectin, an antiparasitic drug for cattle treatment, had an effect on dung-degrading insects, and there was a delay in the pats' deterioration following cow treatment.

5.11 ECOTOXICOLOGICAL EFFECTS

Pharmaceuticals are designed to target particular chemical and metabolic mechanisms in humans and animals, but they often have important side effects as well. Thus far, ecotoxicity testing provided indications of acute effects in vivo in organisms of different trophic levels after short-term exposure and only rarely after long-term (chronic) exposures (Fent et al., 2006). Contemporary literature about the ecotoxicological effects of human pharmaceuticals deals mainly with acute toxicity in standardized tests and is generally focused on aquatic organisms. Moreover, the effects of drug metabolites have rarely been investigated. Photo-transformation products of naproxen, for instance, showed higher toxicities than the parent compound, while genotoxicity was not found (Isidori et al., 2007). Fick et al. (2009) suggested that it is also evident that the levels of fluoroquinolones measured will have ecotoxicological effects, particularly on microbial ecosystems. As bacteria play important roles in the cycling of energy and nutrients, effects on microbial ecology may indirectly have unanticipated consequences for other parts of the ecosystems. The growth of frog tadpoles, for example, exposed to the effluent diluted at a ratio of 1:500 was strongly impaired (Carlsson et al., 2009).

5.11.1 Acute Effects

Acute toxicity data of pharmaceuticals were studied by Webb (2001), who provided a list of about 100 human pharmaceuticals from different sources. By comparing different trophic levels, the researcher suggested that algae were more sensitive to the listed pharmaceuticals than Daphnia magna, and fish were even less sensitive. In the attempt to compare the different classes of pharmaceuticals in terms of acute toxicity, the author noted that the most toxic classes were antidepressants, antibacterials, and antipsychotics (Webb, 2001). The US Geological Survey took the first national look at pharmaceuticals and other compounds within the water. This study took place in 1999–2000 and found organic wastewater contaminants in 80% of the streams they examined (Kolpin et al., 2002). In another survey, fluoxetine (antidepressant, Prozac) and its metabolite were tested for their effects on bivalves. This investigation was carried out because of the rates found within streams and sewage effluent, 0.012 µg/L and 0.099 µg/L respectively (Fong and Molnar, 2008), so concerning fluoxetine and its metabolite norfluoxetine, reproductive behaviors of bivalves are affected. At certain concentrations, reproduction is induced within the bivalve, which can result in the wrong reproductive periods, and therefore, effects upon the larvae and juveniles as food and conditions can reduce survival rates. Another study looked at intersex, the presence of both female and male reproductive characteristics, within bass fish, and these reproductive effects have previously been linked to endocrine active compounds which include some forms of pharmaceuticals.

5.12 POSSIBLE TECHNIQUES TO ABATE PHARMACEUTICAL AND COSMETIC WASTES

To discard pharmaceuticals and cosmetic wastes from different sources without damaging the environment, the wastes must be treated thoroughly to avoid ecological effects on the natural environment like lakes, soil, ocean, groundwater, etc. (Hernandez et al., 2011). The different methods used to remove and treat pharmaceutical and cosmetics wastes from surface water are grouped into two; they are conventional and non-conventional methods (Khan et al., 2004).

5.12.1 Conventional Methods

Conventional methods for eliminating pharmaceutical and cosmetic wastes are activated sludge, rotating biological contactors, and trickling filter methods. Trickling filters and rotating biological contactors are temperature sensitive. They eliminate little biochemical oxygen demand (BOD), and trickling filters require more cost to construct compared to activated sludge systems. Activated sludge methods are more costly to function because more energy is involved for blowers and pumps (NPTEL, 2010).

5.12.1.1 Activated Sludge

Activated sludge is a method that requires a higher number of microorganisms like bacteria, protozoa, and fungi which is aimed at removing organic matter from wastewater (Rajasulochana and Preethy, 2016). This method is a biological treatment process that involves a group assembly of organisms to remove suspended solids and biochemical oxygen demand (BOD). It relies on the rule that wastewater air circulation forms flocs of microorganisms that reduce organic matter and can be removed by sedimentation. This method involves the aeration and settling tanks with other accessories: reoccurrence and waste pumps, mixers and blowers for air circulation, and the device for flow measurement (Liu et al., 2011).

5.12.1.2 Trickling Filter

A trickling filter is a continuous process in which microorganisms responsible for treatment are linked to a dormant pressing material. It consists of a circular tank loaded up with a bearer material. Wastewater is delivered from above and trickles through filter media, permitting organic material in the wastewater to be absorbed by microbial populations like algae, aerobic, anaerobic as well as facultative bacteria. Protozoa and fungi are fixed to the stream as a biological layer. Trickling filters are a competent method in which waste eminence like suspended solids and BOD elimination is high. The method is not complicated when compared with the activated sludge method, but the maintenance and operation are highly skilled since it uses more electric power. Also, skilled labor is essential to preserve the trickling filter processing and operating without any trouble like preventing clogging, ensuring adequate flushing, and controlling filter flies. This process is appropriate for certain comparatively prosperous, heavily populous regions that have a centralized wastewater treatment and better waste system. It is also appropriate for the treatment of grey water. Furthermore, it needs more space contrasted with specific other technologies and has impending for filter flies and odor.

5.12.1.3 Rotating Biological Contactors

The rotating biological contactor (RBC) is a biological treatment technique for organic wastewater treatment, which associates benefits of biological fixed-film like "short hydraulic retention time", "high biomass concentration", "low energy cost", "easy operation", and "insensitivity to toxic substance shock loads". Thus, the aerobic RBC reactor is broadly used for industrial and domestic wastewater treatment (Marcos et al., 2012).

5.12.1.4 Membrane Bioreactors

Membrane bioreactor is a procedure that involves more than one treatment stage. Membrane bioreactor (MBR) techniques are sensitive and combine aerobic- and anoxic-biological treatment with an incorporated membrane organism that may be utilized with most suspended-growth, biological wastewater-treatment systems (Judd, 2008).

5.12.2 Non-conventional Methods

Non-conventional techniques require less sophisticated technology systems, are less expensive, and are easy to maintain and operate. These processes are land-concentrated, unlike the conventional biological methods. They are frequently more effective in eliminating pathogens and do as such dependably and ceaselessly if the method is appropriately planned and not overburdened (Mishra et al., 2016). The non-conventional methods are waste stabilization ponds, constructed wetlands, and oxidation ditch.

5.12.2.1 Waste Stabilization Ponds

The waste stabilization ponds (WSP) technique is a synthetic method that requires shallow basins that include either single or many anaerobic facultative growth ponds used for the treatment of wastewater. The wastewater processes as ingredients are detached by sedimentation or converted by biological and chemical procedures

(Mishra et al., 2016). For facultative ponds, organic matter is broken down further to nitrogen, carbon dioxide, and phosphorous using oxygen created through algae in the pond. This technique improves the quality of wastewater with minimum cost, is simple, and is better for pathogen elimination.

5.12.2.2 Constructed Wetlands

These are strategic techniques that are planned and fabricated to use wetland vegetation to help with the treatment of wastewater in a more meticulous atmosphere than those achieved in natural wetlands. CWs are an ecological and appropriate substitute for secondary and tertiary treatment of industrial and municipal wastewater (Gorito et al., 2018). Also, CWs are appropriate for the elimination of suspended solids, organic materials, nutrients, heavy metals, pathogens, and toxic pollutants. However, CWs are not ideal for the raw waste treatment and pretreatment of industrial wastewater to preserve the biological equilibrium of the wetland ecology.

5.12.2.3 Oxidation Ditches (OD)

Usual OD treatment techniques comprise a single or multichannel arrangement within a ring or oval. Oxidation ditches are characteristically comprehensive mix techniques; however, they can be amended. Preliminary treatment like bar screens and grit elimination usually leads to the oxidation ditch. Other methods used to abate pharmaceutical and cosmetic waste from point and non-point sources are phytoremediation, bioremediation (bacteria treatment), and mycoremediation (fungi treatment).

5.13 PHYTOREMEDIATION

Phytoremediation is an effective and economical method of treating organic and inorganic waste across the globe. It makes use of the natural phenomenon of degradation and accumulation of green plants as a competent agent for the treatment. Plants have an exceptional ability to treat the pharmaceuticals from wastewater and soil, especially *Typha latifolia* and *Phragmites karka*. Also, it was reported that phenol generated from industrial wastewater is phytoremediation with the aid of peroxidases of tomato hairy root cultures (Gonzalez et al., 2006). Phytoremediation can be an option for the removal of these compounds (pharmaceuticals and cosmetic wastes) from various environmental matrices. Active plant processes, including uptake and metabolism, can remove these compounds in wetland systems. Duckweeds (*Lemna minor*) have been found efficient in the uptake and metabolization of chlorinated and fluorinated phenols (González et al., 2006).

5.13.1 Bioremediation (Removal of Contaminants by Bacterial Treatment)

Bioremediation is the use of bacteria for the treatment of pollutants or waste. It is an effective method used for the treatment of pharmaceutical waste. This can be achieved with a naturally resistant microbial strain that can transform the toxic form into a less toxic form (Gorito et al., 2018). This method can be made more successful by adding nutrients as stimulants for biodegrading microorganisms, known as biostimulation. This process remits enough degradation and transformation products; the influence on the environment must be a matter of concern. Some notable examples of microbes associated with the degradation of pharmaceuticals are *Bacillus subtilis*, *Cyanobacterium phormidium*, and *Bacillus cereus*. Other abundant groups include Proteobacteria, Bacteroidetes, and Actinobacteria as well as the bacterial groups Acidobacteria (Gorito et al., 2018).

5.13.2 Mycoremediation (Removal of Pharmaceutical Compounds by Fungal Treatment)

Fungi are also an agent for remediating various pollutants, including pharmaceuticals and cosmetics, from surface water. The fungus *Trametes versicolor* is associated with the recalcitrant analgesics, anti-inflammatories, antibiotics, and psychotropic drugs. *Candida tropicalis*, *Debaryomyces polymorphus*, and *Candida zeylanoides* utilized and eliminated synthetic dyes. *Aspergillus oryzae*, *Aspergillus variabilis*, and *Tolypothrix ceytonica* removed organic matter, copper, and zinc. Fungi genera is an efficient tool for remediating waste from pharmaceutical industries; the only deficiency is the slow growth rate and formation of filaments and spores. Various fungal species such as *Aspergillus niger*, *Aspergillus fumigatus*, *Aspergillus niveus*, *Penicillium decumbens*, and *Penicillium lignorum* also are used for the reduction of pharmaceuticals from wastes (Angayarkanni et al., 2003). Features of some organisms in surface water are shown in Table 5.1.

5.14 SUMMARY

The plankton is an important group of organisms because they fix carbon dioxide and produce oxygen. The continuous need for pharmaceuticals in the health sector has led to aquatic pollution. Cosmetics and pharmaceutical compounds remain active after being discharged into the environment so they can affect the aquatic organisms by interfering with their biological system.

Antibiotics, hormones, analgesics, antidepressants, and antineoplastics are pharmaceutical products that pose high environmental risks. These compounds are distributed to the environment through the point and diffuse sources. The point sources originate from pharmaceutical manufacturing plants, healthcare institutions, external care facilities, personal care products, and veterinary offices. Pharmaceutical wastes are grouped into solid and liquid waste, while cosmetics wastes arise from those materials used for packaging in the beauty industry.

5.15 CONCLUSION

The presence of cosmetics and pharmaceutical waste in aquatic environments especially in plankton communities cannot be efficiently emphasized due to the increase in the used of pharmaceuticals and cosmetics products. Hence, the detrimental effects on the aquatic environment can be eliminated by proper channeling of pharmaceutical and cosmetic by-products.

TABLE 5.1
Some Organisms in Surface Water.

S/No	Organism(s)	Features	Examples
1.	Bacteria	They are microscopic organisms that are unicellular, prokaryotic, and lack membrane bound organelles.	*Staphylococcus aureus, Escherichia coli, Vibrio cholera, Pseudomonas aeruginosa, Klebsiella pneumonia*
2.	Protozoa	Protozoa are unicellular, eukaryotic microorganisms lacking a cell wall; their nucleus is enclosed in a membrane.	copepods, Daphnia, Radiolarians, Foraminiferans
3.	Archaea	Like bacteria, archaea lack interior membranes and organelles. They survive in extreme environments. Archaea have rigid cell walls with diverse structures.	Methanobacteriales, Methanosarcinales, Methanomicrobiales, and Nitrosopumilaceae
4.	Viruses	Viruses are obligate intracellular parasites and contain either DNA or RNA but not both. They also contain protective protein coating called the capsid and are either single or double stranded.	enterovirus, norovirus, hepatitis A virus, coronaviruses
5.	Fungi	They are eukaryotes, may be unicellular or multicellular. Fungi lack chlorophyll and, hence, are heterotrophic.	*Fusarium oxysporum, Exophiala dermatitidis, Epicoccum nigrum*
6.	Helminths	Helminths are invertebrates, eukaryotes, multicellular, and parasitic animals. They have bilateral symmetry.	Schistosomes, Ascaris, Echinococcus, Enterobius, Trichuris
7.	Algae	Algae can be unicellular or multicellular. They are photosynthetic organisms and are found mostly in moist environment.	chlorella, cryptomonads, dinoflagellates, diatoms
8.	Crustaceans	Crustaceans are hard with flexible exoskeleton. They have two compound eyes and two pairs of maxillae on their heads.	crayfish, prawns, crabs, shrimps, copepods, ostracoda
9.	Mollusks	The bodies of mollusks are made up of three main parts, including a muscular foot, a visceral mass, and a mantle. The foot is an appendage used for either burrowing or movement.	corbicula, abalones, winkles, conchs, periwinkles, oysters
10.	Coelenterates	Coelenterates are multicellular organisms. They are diploblastic, with two layers of cells, an outer layer called the ectoderm and the inner layer called the endoderm.	hydra, sea pens, comb jellies, obelia, aurelia, sea anemones

5.16 RECOMMENDATION

- It is recommended that health professionals should educate and create awareness in the community on discarding unused, expired drugs and other wastes from cosmetics.

- Hospitals and the pharmaceutical industry should be able to manage their hazardous wastes through the implementation of waste programs that are supported by appropriate legislation, standards, and controls.

REFERENCES

Angayarkanni, J., Palaniswamy, M., and Swaminathan, K. (2003). Biotreatment of distillery effluent using Aspergillusniveus. *Bulletin of Environmental Contamination and Toxicology*, 70, 268–277.

Brodin, T., Fick, J., Jonsson, M., and Klaminder, J. (2013). Dilute concentrations of a psychiatric drug alter behavior of fish from natural populations. *Science*, 339(6121), 814–815.

Carlsson, G., Örn, S., and Larsson, D.G. (2009). Effluent from bulk drug production is toxic to aquatic vertebrates. *Environmental Toxicology and Chemistry*, 28(12), 2656–2662.

Cleuvers, M. (2008). Chronic mixture toxicity of pharmaceuticals to Daphnia–the example of nonsteroidal anti-inflammatory drugs. In *Pharmaceuticals in the Environment*, Springer, Berlin, Heidelberg, pp. 277–284.

Cooper, E.R., Siewicki, T.C., and Phillips, K. (2008). Preliminary risk assessment database and risk ranking of pharmaceuticals in the environment. *Science of the Total Environment*, 398(1), 26–33.

Debenest, T., Silvestre, J., Coste, M., and Pinelli, E. (2010). Effects of pesticides on freshwater diatoms. *Reviews of Environmental Contamination and Toxicology*, 203, 87–103.

DellaGreca, M., Fiorentino, A., Iesce, M., Isidori, M., Nardelli, A., and Previtera, L. (2003). Identification of phototransformation products of prednisone by sunlight. Toxicity of the drug and its derivatives on aquatic organisms. *Environmental Toxicology and Chemistry*, 22, 534–539.

Fent, K., Weston, A., and Caminada, D. (2006). Review: Ecotoxicology of human pharmaceuticals. *Aquatic Toxicology*, 76, 122–159.

Fick, J., Söderström, H., Lindberg, R.H., Phan, C., Tysklind, M., and Larsson, D.G.J. (2009). Contamination of surface, ground, and drinking water from pharmaceutical production. *Environmental Toxicology and Chemistry*, 28, 2522–2527.

Fong, P.P., and Molnar, N. (2008). Norfluoxetine induces spawning and parturition in estuarine and freshwater bivalves. *Bulletin of Environmental Contamination and Toxicology*, 81(6), 535.

Geiger, E., Hornek-Gausterer, R., and Saçan, M.T. (2016). Single and mixture toxicity of pharmaceuticals and chlorophenols to freshwater algae *Chlorella vulgaris*. *Ecotoxicology and Environmental Safety*, 129, 189–198.

Germain, R.M., Mayfield, M.M., and Gilbert, B. (2018). The 'filtering' metaphor revisited: Competition and environment jointly structure invasibility and coexistence. *Biology Letters*, 14(8), 20180460.

González, S., Petrović, M., and Barceló, D. (2006). Advanced liquid chromatography/mass spectrometry (LC-MS) methods applied to wastewater removal and the fate of surfactants in the environment. *Trends in Analytical Chemistry*, 26, 116–124.

Gorito, A.M., Ribeiro, A.R., Gomes, C.R., Almeida, C.M.R., and Silva, A.M.T. (2018). Constructed wetland microcosms for the removal of organic micropollutants from freshwater aquaculture effluents. *Science of the Total Environment*, 644, 1171–1180.

Goudard, A., and Loreau, M. (2008). Nontrophic interactions, biodiversity, and ecosystem functioning: An interaction web model. *American Naturalist*, 171, 91–106.

Guardabassi, L., Petersen, A., Olsen, J.E., and Dalsgaard, A. (2021). Antibiotic resistance in Acinetobacterspp. Isolated from sewers receiving waste effluent from a hospital and a pharmaceutical plant. *Applied and Environmental Microbiology*, 64(9), 3499–3502.

Gurudatta, S., Anubhuti, S., Priyanka, S., and Akanksha, G., (2020). Sources, fate and impact of pharmaceutical and personal care products in the environment and their different treatment technologies. In *Microbe Mediated Remediation of Environmental Contaminants*. Woodhead Publishing series in food science, Technology and Nutrition Elsevier, pp. 391–407. DOI: 10.1016/B978-0-12-821199-1.00029-8.

Hahn, T., and Schulz, R. (2007). Indirect effects of antibiotics in the aquatic environment: A laboratory study on detritivore food selection behavior. *Human and Ecological Risk Assessment*, 13(3), 535–542.

Hai Nguyen, V., and Reinhard, M. (2010). Emerging contaminants: A potential environmental risk in sewage sludge amendment to agricultural soils. *Environmental Pollution*, 158(5), 1643–1649.

Halling-Sbrensen, B. (2000). Algal toxicity of antibacterial agents used in intensive farming. *Chemosphere*, 40, 731–739.

Halling-Sbrensen, B., Nor Nielsen, S., Lanzky, P., Ingerslev, F., HoltenLqtzhoft, H., and Jbrgensen, S. (1998). Occurrence, fate and effects of pharmaceutical substances in the environment a review. *Chemosphere*, 36, 357–393.

Harrass, M.C., Kindig, A.C., and Taub, F.B. (2018). Responses of blue-green and green algae to streptomycin in unialgal and paired culture. *Aquatic Toxicology*, 6(1), 1–11.

Heberer, T. (2002). Occurrence, fate, and removal of pharmaceutical residues in the aquatic environment: A review of recent research data. *Toxicology Letters*, 131, 5–17.

Hernandez, L.L., Temmink, H., Zeeman, G., and Buisman, C.J.N. (2011). Characterization and anaerobic biodegradability of grey water. *Desalination*, 270, 111–115.

Isidori, M., Lavorgna, M., Nardelli, A., Parrella, A., Previtera, L., and Rubino, M. (2005). Ecotoxicity of naproxen and its phototransformation products. *Science of the Total Environment*, 348, 93–101.

Isidori, M., Nardelli, A., Pascarella, L., Rubino, M., and Parrella, A. (2007). Toxic and genotoxicimpact of fibrates and their photoproducts on non-target organisms. *Environment International*, 33, 635–641.

Judd, S. (2008). The status of membrane bioreactor technology. *Trends in Biotechnology*, 26(2), 109–116.

Khan, N., Ibrahim, A., and Subramaniam, P. (2004). Elimination of heavy metals from wastewater using agricultural wastes as adsorbents. *Malaysian Journal of Science*, 23, 43–51.

Kidd, K.A., Blanchfield, P.J., Mills, K.H., Palace, V.P., Evans, R.E., Lazorchak, J.M., and Flick, R.W. (2007). Collapse of a fish population after exposure to a synthetic estrogen. *Proceedings of the National Academy of Sciences*, 104(21), 8897–8901.

Kolpin, D., Furlong, E., Meyer, M., Thurman, E., Zaugg, S., and Barber, L. (2002). Pharmaceuticals, hormones, and other organic wastewater contaminants in US streams, A national reconnaissance. *Environmental Science and Technology*, 36, 1202–1211.

Kümmerer, K. (2001). *Pharmaceuticals in the Environment: Sources, Fate Effects and Risks*, Springer, Berlin.

Küster, A., and Adler, N. (2014). Pharmaceuticals in the environment: Scientific evidence of risks and its regulation Philos. *Transactions of the Royal Society B*, 369(1656), 20130587.

Lagesson, A., Fahlman, J., Brodin, T., Fick, J., Jonsson, M., Byström, P., and Klaminder, J. (2016). Bioaccumulation of five pharmaceuticals at multiple trophic levels in an aquatic food web-Insights from a field experiment. *Science of the Total Environment*, 568, 208–215.

Laxminarayan, R., Duse, A., and Wattal, C., (2013). Broad-spectrum antibiotics and the continuing threat of resistance. *The Lancet Infectious Diseases*, 13(12), 1057–1098.

Lee, W.Y., and Bird, C.R. (2021). Chronic toxicity of ocean-dumped pharmaceutical wastes to the marine amphipod Amphithoevalida. *Marine Pollution Bulletin*, 14(4), 150–153.

Liu, J.J., Wang, X.C., and Fan, B. (2011). Characteristics of PAHs adsorption on inorganic particles and activated sludge in domestic wastewater treatment. *Bioresource Technology*, 102(9), 5305–5311.

Lumaret, J.P. (2016). Toxicité de certainshelminthicides vis-à-vis des insectescoprophagesetconséquencessur la disparition des excréments de la surface du sol. Actaoecologica. *Oecologiaapplicata*, 7(4), 313–324.

Macri, A., Stazi, A.V., and Di Delupis, G.D. (2018). Acute toxicity of furazolidone on Artemia salina, Daphnia magna, and Culex pipiens molestus larvae. *Ecotoxicology and Environmental Safety*, 16(2), 90–94.

Marcos, R.V., Gilberto, C.B.M., and Márcio, R.V.N. (2012). Wastewater treatment in trickling filters using LUFFA cyllindrica as biofilm supporting medium. *Journal of Urban and Environmental Engineering*, 6(2), 57–66.

Martel, A., Meunier, L., and Crepineau, F. (2021). Antibiotics in animal production. The regulatory perspective. *Annual Review of Animal Biosciences*, 9, 445–462.

Mishra, V.K., Upadhyay, A.R., and Pandey S.K. (2016). Phytoremediation of waste water: A review on the use of plants to treat waste water. *Environmental Science and Pollution Research International*, 23(22), 22411–22427.

National Programme on Technology Enhanced Learning (NPTEL). (2010). *Wastewater Treatment*. Course Notes. www.nptel. iitm.ac.in/courses/Webcourse-contents/IIT

Ngqwala, N.P., and Muchesa, P. (2020). Occurrence of pharmaceuticals in aquatic environments: A review and potential impacts in South Africa. *South African Journal of Science*, 116(7–8), 1–7.

Nicol, G.D., and Miller, W.H. (2020). Cyclic GMP injected into retinal rod outer segments increases latency and amplitude of response to illumination. *Proceedings of the National Academy of Sciences*, 75(10), 5217–5220.

Oaks, J.L., Gilbert, M., Virani, M.Z., Watson, R.T., Meteyer, C.U., Rideout, B.A., Shivaprasad, H.L., Ahmed, S., Chudhry, M.J.I., Arshad, M., Mahmood, S., Ali, A., and Khan, A.A. (2010). Diclofenac residues as the cause of vulture population decline in Pakistan. *Nature*, 427, 630–633.

Pierella, K., Juan, J., Ibarbulz, F.M., and Bowler, C. (2020). Exploration of Marine Phytoplankton from their historical appreciation to the Omics era. *Journal of Plankton Research*, 42, 595–612.

Rajasulochana, P., and Preethy, V. (2016). Comparison on efficiency of various techniques in treatment of waste and sewage water—A comprehensive review. *Resource-Efficient Technologies*, 2, 175–184.

Reynolds, J. (2019). *Sodium Cromoglycate and Related Anti-Allergic Agents*, Pharmaceutical Press, London.

Sanyal, A.K., Roy, D., Chowdhury, B., and Banerjee, A.B. (2016). Ibuprofen, a unique anti-inflammatory compound with antifungal activity against dermatophytes. *Letters in Applied Microbiology*, 17(3), 109–111.

Sherer, J.T. (2006). Pharmaceuticals in the environment. *American Journal of Health-System Pharmacy*, 63(2), 174–178.

Smith, D.J. (2013). Aeroplankton and the need for a global monitoring network. *Bioscience*, 63(7), 515–516.

Smith, J.D., and Johnson, A.B. (2023). Classification and mechanisms of action of antibiotics. *Journal of Antibiotic Research*, 15(3), 123–145.

Taggart, M.A., Cuthbert, R., Das, D., Sashikumar, C., Pain, D.J., Green, R.E., Feltrer, Y., Shultz, S., Cunningham, A.A., and Meharg, A.A. (2015). Diclofenac disposition in Indian cow and goat with reference to Gyps vulture population declines. *Environmental Pollution*, 147(1), 60–65.

Ternes, T. (1998). Occurrence of drugs in German sewage treatments plants and rivers. *Water Research*, 32, 3245–3260.

Wall, R., and Strong, L. (2017). Environmental consequences of treating cattle with the anti-parasitic drug ivermectin. *Nature*, 327(6121), 418–421.

Webb, S.F. (2001). A data based perspective on the environmental risk assessment of human pharmaceuticals II—Aquatic risk characterisation. In *Pharmaceuticals in the Environment*, Springer, Berlin, Heidelberg, pp. 203–219.

Weston, D.C., and Caminada, T. (2006). Ecotoxicology of human pharmaceuticals. *Aquatic Toxicology*, 76, 122–159.

Xin, X., Huang, G., and Zhang, B. (2020). Review of aquatic toxicity of pharmaceuticals and personal care products to alga. *Journal of Hazardous Materials*, 410(7), 124619.

Zhang, W.J., Pan, Y.B., Yang, J., Chen, H.H., Holohan, B., and Vaudrey, J. (2018). The diversity and biogeography of abundant and rare intertidal marine microeukaryotes explained by environment and limitation. *Environmental Microbiology*, 20, 462–476.

Zuccato, E, Castiglioni, S., Fanelli, R., Reitano, G., Bagnati, R., Chiabrando, C., Pomati, F., Rossetti, C., and Calamari, D. (2006). Pharmaceuticals in the environment in Italy: Causes, occurrence, effects and control. *Environmental Science and Pollution Research*, 13, 15–21.

6 Influence of Cosmetic and Pharmaceutical Wastes on Community of Plankton

*K.S. Shreenidhi, B. Vijaya Geetha,
Sree S. Sai Sandhya, S. Jothi Murugan, and
G. Sarada*

6.1 INTRODUCTION

6.1.1 Exploring the World of Plankton: Nature's Drifting Microcosms

Plankton are a broad group of organisms that live in aquatic environments and are unable to move against the current or waves. Plankton are generally microscopic, often less than one inch in length, but also include larger species like some crustaceans and jellyfish.

The two groups of the general plankton community are as follows:

a. Plants, or phytoplankton.
b. Animals, or zooplankton.

6.1.1.1 Photosynthetic Powerhouses: Understanding Phytoplankton's Vital Role in Marine Ecosystems

Microorganisms known as phytoplankton are an important part of the marine food chain (Sommer, 2002). They carry out photosynthesis, which allows them to absorb carbon dioxide and release oxygen, and they also take in sunlight. Sunlight is necessary for their development (Ajani et al., 2020; Mackas et al., 2007).

6.1.1.2 Unveiling Zooplankton: Microscopic Marvels Shaping Marine Food Webs

Zooplankton, including microscopic animals such as krill, sea snails, pelagic worms, etc., the young of larger invertebrates and fish, and weak swimmers like jellyfish are consumed by larger animals (Istu Septania, 2021). To avoid being eaten, zooplankton typically drifts in deeper waters during the day and rises to the surface at night to feed on phytoplankton (Chiba et al., 2015). The size of zooplankton is especially significant, as it serves as a marker for various biological and ecological characteristics, including metabolism, feeding tactics (Kiørboe, 2011), and food chain connections (Barton et al., 2013).

DOI: 10.1201/9781003362975-6

Plankton is extremely important to the aquatic ecosystem and very sensitive to environmental changes, such as temperature, salinity, pH level, and nutrient concentration of the water (Jernberg et al., 2017). Excessive nutrients, often from sources like agricultural runoff, sewage, or industrial waste, can lead to an overabundance of certain types of plankton, such as algae. When these nutrients, particularly nitrogen and phosphorus, are present in high concentrations in water bodies like lakes, rivers, or coastal areas, they act as a kind of fertilizer for these microscopic organisms. Harmful algal blooms (HABs) and plankton blooms, like red tides, are the result. Variations in the quantity of phytoplankton can significantly affect the populations of zooplankton, which in turn affects other species in the food chain because many zooplankton rely on them for food. We are going to discuss various factors that affect the plankton community (Costello et al., 2010).

Physiological parameters are temperature, light, pH, and partial pressures of the required gases such as CO_2. Like terrestrial plants, phytoplankton also require nitrogen, calcium, potassium, etc., in various amounts (Chiba et al., 2018). A drastic decline in the phytoplankton biomass can be seen in ocean regions, where the temperature rises above normal, implying that climate change plays a huge role in the growth of phytoplankton (Weber, 1968). Loss of the phytoplankton community causes a huge imbalance in the aquatic ecosystem since every organism in the waters either feeds on the phytoplankton or feeds on the others that depend directly on it.

Human activities such as disposing of improperly recycled effluents into the oceans and oil spillage result in the increased concentrations of heavy metals, persistent organic pollutants, etc.

While nitrogen and potassium are vital for their growth, an abundance of these inorganic nutrients can cause over-blooming of the plankton forming HABs (Tweddle et al., 2018). An increase in the concentration of such nutrients occurs due to agricultural runoff, pharmaceutical and detergent industries' effluents, etc. Fertilizers that help improve the growth of crops on the one hand also affect the plankton communities indirectly (Berthold & Campbell, 2021). These HABs pollute their habitats by producing toxic compounds that have direct and indirect harmful effects on fish, shellfish, mammals, birds, and humans. Harmful algal blooms of cyanobacteria do not necessarily affect humans and animals by infecting them but rather, by growing too dense or producing toxins such as microcystin, cylindrospermopsin, anatoxin, etc. (Beiras & Tato, 2019).

HABs of dinoflagellates or diatoms are called red tides, as they make the water appear red. Cosmetic and pharmaceutical wastes (CnPWs) release toxins like brevetoxin, azaspiracid, and domoic acid, posing risks to humans and coexisting organisms (Duarte et al., 2021). In our exploration of CnPWs, we'll delve into their broader impact on the aquatic ecosystem. We aim to unravel and understand the intricate dynamics, contributing to informed conservation strategies (Jo et al., 2021).

6.2 PHARMACEUTICAL WASTES VS PLANKTON COMMUNITY

6.2.1 Effect of Pharmaceutical Industries on the General Plankton Community

Pharmaceutical wastes mainly consist of pharmaceutical and personal care products (PPCPs). Used and unused medicines, expired prescription pharmaceuticals, personal care products, and over-the-counter medications have increased since the development of standard medical waste regulations as being a major public and environmental health concern (Desforges et al., 2015).

The primary aim of pharmaceutical industries is to enhance human health, yet a healthy planet is essential for fostering people's well-being. The interconnection between our planet and human life is inherent (Stone et al., 2023). Pharma must stick to its core vision and act now to reduce the emission of pollutants and play its part in reaching global net zero goals. Quantification and treatment of pharmaceutical effluents is challenging since they contain high amounts of organic matter (alcohols, acetone), molecules (aromatic compounds, chlorinated hydrocarbons), and other heavy metals (Alfonina & Tashlykova, 2018). The discharge of flora and fauna affects the marine ecosystem and overall food chain of the environment. They also contribute to cytotoxic and genotoxic effects in aquatic environments.

6.2.2 Active Pharmaceutical Ingredients (APIs)

The pharmaceutical products and their degradation products are introduced into the environment via the following:

1. Discharges from pharmaceutical industries.
2. Hospitals.
3. Agricultural sources.
4. Wastewater treatment plant effluents.
5. Mammalian excretion of unmetabolized, partially metabolized, and fully metabolized forms of drugs.

Collectively, pharmaceuticals and their partially and fully metabolized forms are called active pharmaceutical ingredients (APIs) (Chia et al., 2021).

Experimental evidence suggests that APIs have unfavorable effects on the general phytoplankton community, which indirectly affects higher trophic-level organisms.

The pharmaceutical affects the growth, photosynthesis, metabolism, and community structure of plankton, thereby affecting the biogeochemical cycle of nutrients (Kim et al., 2022)

The impact of pharmaceuticals on the morphological and physiological characteristics of phytoplankton can be observed at the individual and communal levels of organization. Alterations in cell size, form, and bio-volume have been the most prominent changes that have been observed. Phytoplankton cell size responds to the discharge of AIPs in environmental conditions (Borics et al., 2020) and is also a factor determining the function and structure of the plankton community.

6.2.3 Microplastics

Microplastics are tiny plastic particles that result from both commercial product development and the breakdown of larger plastics (Mangolte et al., 2022). Officially, they are defined as plastics less than 5 mm. Microplastics from medicinal products are released into the wastewater through pharmaceutical manufacturing activities and improper disposal of unused medicine by the patients. As a pollutant, microplastics can be harmful to the environment and animal health (Sharp et al., 2020).

There are four main types of microplastics:

1. Fragments.
2. Nurdles.
3. Microbeads.
4. Microfibers.

Microplastic fragments tiny pieces that break off from larger items, like a plastic bottle cap. The second category is nurdles. They are small plastic beads that manufacturers produce and sell to other companies to melt down into less fibers, but that's still a lot! It's no wonder microfibers are one of the biggest contributors to microplastic pollution (Case et al., 2008).

Planktons are likely to be increasingly affected as the introduced and degraded MPs particle sizes are similar to their prey size (zooplankton and ichthyoplankton) or their ability to entrain particles (phytoplankton). Marine organisms can be exposed through direct or indirect ingestion of microplastics through prey items or using respiration. Through either of the pathways, microplastics intake can result in unfavorable physical and chemical impacts on marine organisms (Bottrell et al., 2019). Exposure to microplastics in aquatic environments results in adverse phenomena such as bio-items we buy at the store (e.g. "raw material" that makes your toothbrush). These microplastics can be released into the ocean when a shipping container on a cargo ship goes overboard.

The third type of microplastics is one you might be familiar with—microbeads. These are tiny plastic balls that get incorporated into our personal care products as exfoliants. You can avoid these microplastics by checking for "polypropylene" and "polyethylene" on the ingredients list when buying toothpaste and face wash.

The fourth category is microfibers. When we wash clothes made of synthetic fibers, thousands of tiny plastic threads are swept into the wastewater stream, eventually flushing out into the ocean. Acrylic fabrics are the most notorious, releasing over 700,000 fibers on their first wash. Subsequent washes release cumulation and biomagnification (Hitchcock, 2022).

Bioaccumulation (or body burden) is defined as the net uptake of a contaminant (microplastics) from the environment in all possible ways (e.g. contact, ingestion, respiration) from any source (e.g. water, sediment, prey). In other words, bioaccumulation occurs when the uptake of a contaminant is greater than the ability of an organism to ingest a contaminant. Potential impacts of bioaccumulation include physical retention of MPs in digestive tracts and chemical leaching of plastic additives into tissues.

Biomagnification across a food web can thus be defined as the increase in the concentration of a contaminant (microplastics) in one organism compared to the concentration in its prey. This biomagnification indirectly affects human beings by entering into our daily food practice using fish. This leads to prostate cancer, infertility, and cognitive impairments. Apart from this, the intake of microplastics causes a direct effect on the plankton community and decreases feeding on algal prey due to microplastic ingestion, which has been seen in an experiment (see Figure 6.1). It

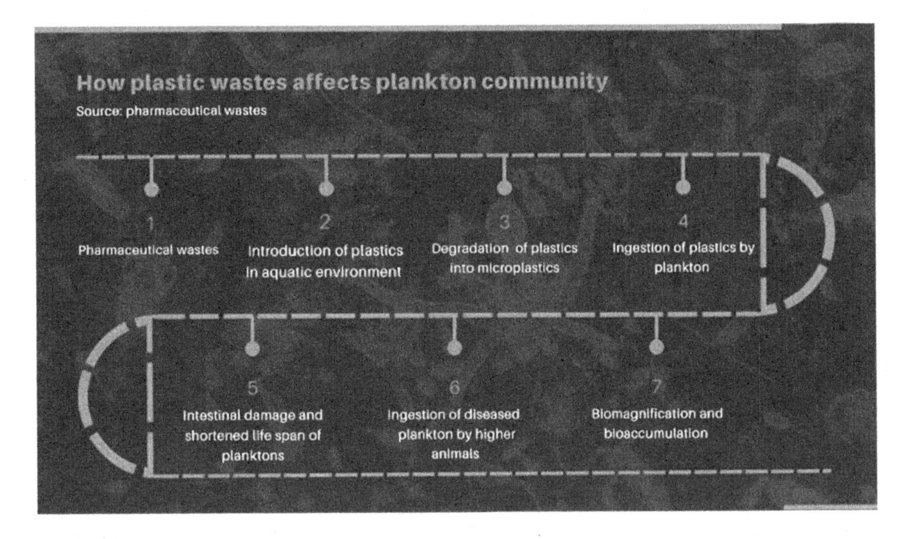

FIGURE 6.1 Effect of pharmaceutical wastes on plankton community and to the environment (OWN).

also disrupts physical development as seen in the planktotrophic pluteus larvae of the mammal. The sea urchin *Paracentrotus lividus* developed an altered pluteus shape when microplastics were ingested. Inhalation of microplastics through air, fibers, or aerosols at the water-air interface aquatic environment is also an important route of exposure for air-breathing marine mammals.

The hydrophobic nature of microplastics leads to the formation of aggregations that lead to obstruction that further reduces motility, ingestion, reproduction, and mechano-reception, by hindering externally on the appendages, swimming legs, feeding apparatus, antennae, and furca of copepods. Therefore, microplastics have become a serious issue to be sorted out, as they contribute to most of the adverse effects on the plankton community, thereby affecting the entire food web of the planet.

6.2.4 Plastics

Plankton are photosynthetic organisms. Plastic blocks sunlight from reaching plankton, which prevents the photosynthesizing of the organism and thereby oxygenating the ocean. Plankton cannot survive without sunlight. This results in dead zones, rapidly increasing in size and number. This disrupts the food chain of the entire ecosystem. They also attach themselves to plastic particles. Plastic acts as a magnet for other contaminants. Since plastics are oily petroleum-based products, they, in turn, attract other petroleum-based chemicals, debris, and oils. These contaminant-carrying globules (plankton, plastic, and added contaminants) are consumed by zooplanktons, fishes, and other mammals, causing higher concentrations of contaminants, all the way up the food chain, wreaking havoc on the health of all these creatures.

6.2.5 Organic and Inorganic Wastes

Dissolved organic matter (DOM) is closely related to phytoplankton blooms, one of the most concerning environmental problems globally. This not only leads to the consumption of excess oxygen and the death of other aquatic organisms but also produces toxins that may cause serious public health issues. They can discolor water, pollute beaches, or cause drinking water and fish to taste foul. Since sunlight is a fundamental need for the growth of plankton, dense blooms disrupt the growth of algae and other plankton by blocking the sunlight from reaching them (Spilling et al., 2019).

The abundance of phytoplankton biomass is linked with the amount of protein-like DOM ($p < 0.01$). TITAN reveals that protein-like DOM gradients generated change points in phytoplankton cell abundance, and canonical correspondence analysis showed that phytoplankton taxonomic composition was significantly changed with protein-like DOM.

Inorganic phosphate in the form of p has a negative correlation with dissolved oxygen (DO). DO is essential for the growth of phytoplankton, without which, the community will be disrupted. Nitrogen is an essential component for the growth of phytoplankton. Available in the form of nitrate is dissolved nitrogen, which influences phytoplankton community structure by favoring the growth of certain taxa over others.

The *Journal of Plankton Research* states *that* phosphate has been found to limit phytoplankton biomass on short timescales in the subtropical North Atlantic due to the high activity of alkaline phosphatase and N:P ratios higher in phosphate-dissolved pools. Marine phytoplankton have diverse phosphate acquisition systems and storage capacities, which affect their response to phosphorus increments.

While the response of phytoplankton to contamination of nutrients has been studied in laboratory and field experiments and biogeochemical models, co-limitation is harder to access. However, vast areas of the ocean are oligotrophic (poor in plant nutrients and containing abundant oxygen in the deeper parts) harboring phytoplankton communities depleted by several nutrients. The decreased primary production depends on cellular requirements and cell size of the phytoplankton taxa, which are related to ratios of the macromolecular pools, reserves, and energy. The addition of macro- and micronutrients increased the abundance of the "velocity-adapted" diatoms in most nutrient treatments. Here are some of the substances that are present in most cosmetic products and their effects on the plankton community (Johns et al., 2001).

6.3 SOURCES OF PHARMACEUTICAL WASTES IN THE MARINE ENVIRONMENT

When pharmaceutical wastes are discharged into the aquatic environment, their metabolites undergo biotic and abiotic transformation or degradation and form suspended particulate matter (SPM) and sediments. SPM also accumulates in the tissues of marine organisms (Sally et al., 2019). Hence, it's necessary to know the mode of introduction of toxins in the marine environment. The major ways through which the toxic wastes from pharmaceutical wastes are as follows:

1. Sewage.
2. Aquaculture.
3. Waste disposal.
4. Discharge of ion-rich water without proper treatment.

6.3.1 Mechanism of Action of Pharmaceuticals on Phytoplankton: Subcellular Endpoints

The general mechanism of degraded growth of plankton by pharmaceutical waste can be observed by monitoring subcellular endpoints such as photosynthetic parameters, oxidative stress response parameters, proteins, global metabolite profiles, and gene expression, which are critical (Goldyn & Kowalczewski-Madura, 2007). This approach emphasizes the need for *Omics* analysis of pharmaceutical exposure-induced cellular changes of phytoplankton (Murphy et al., 2020).

6.3.2 Oxidative Stress Response

Oxidative stress is a phenomenon caused by an imbalance between the production and accumulation of reactive oxygen species (ROS) in cells and tissues and the ability of a biological system to detoxify these reactive products (Florescu et al., 2022).

6.3.3 Global Metabolite

A metabolite profile can be defined empirically as the set of all metabolites or derivative products (identified or unknown) detected by analyzing a sample using a particular analytical technique, together with an estimate of quantity.

6.3.4 Effects of Cosmetics on Plankton Community

A growing awareness of the adverse effects associated with skin diseases due to direct ultraviolet (UV) radiation over the past decades paved the way for the increased use of sunscreen cosmetic products, leading to the introduction of new chemical compounds in the marine ecosystem. Concentrations of chemical UV filters included in the production of sunscreens and other cosmetics, such as benzophenone 3, 4-methylbenzylidene camphor, titanium dioxide, and zinc oxide, are detected in nearshore waters along the day and mainly concentrated in the surface microlayer (Cullor, 2022).

The existence of these compounds in seawater shows adverse effects on phytoplankton. Dissolution of sunscreens in seawater also releases inorganic nutrients (N, P, and Si forms) that lead to phytoplankton blooms (Culliney, 2020; Tovar-Sánchez et al., 2013).

Sun protection cosmetics are composed of the following:

1. Organic UV filters (para-aminobenzoate, cinnamates, benzophenones, dibenzoyl methane, camphor derivatives, and benzimidazoles, which absorb the UV radiations).

2. Inorganic UV chemical filters (i.e. TiO2 and ZnO) reflect and scatter the UV radiation, protecting human skin from the direct radiation of sunlight (Yakubu et al., 2018).
3. Preservatives (e.g. parabens derivatives).
4. Coloring agents (e.g. ammonium sulfate, copper powder, ferric ammonium ferrocyanide, iron and zinc oxides, etc.).
5. Film-forming agents (e.g. acrylates and acrylamides).
6. Surfactants, chelators, and viscosity controllers (e.g. potassium acetyl phosphate, pentasodium methylene diamine tetra methylene phosphonate among others) (Dickinson, 2021).

6.3.4.1 Microbeads

Microbead technology is the most commonly used cutting-edge cosmetic technology (Möhlenkamp et al., 2018). Microbeads are small plastics made from polyethylene. In a shower gel, there is an equivalent amount of combined plastic in the gel itself, as there is plastic in the bottle that holds it. Titanium dioxide is another commonly used chemical ingredient usually found as nanoparticles used in mineral makeup and sunscreens. This chemical has been the reason for damage in snail DNA and reduction in the growth of phytoplankton. This is of huge concern since the phytoplankton community is responsible for providing more than half of our planet's oxygen and is a major source of food for large amounts of higher plants and aquatic animals (Murrell and Lores, 2004). Losing this marine species will lead to enormous decreases in oxygen levels on land and in water while dooming aquatic life as we know it. The loss of marine species would lead to an enormous plummet in available oxygen levels on land and in water, dooming aquatic life and correspondingly affecting terrestrial life.

6.3.4.2 Parabens

Paraben is a common chemical preservative to increase the shelf life of products. In-depth evidence has been collected that proves that paraben is the major cause of hormonal damage in marine animals, harming reproductive organs and, hence, fertility. Parabens have also been proven to play an immense role in depleting the coral, accumulating in the tissues of marine and land animals.

6.3.4.3 Coal Tar

Anthracene, a common polycyclic aromatic hydrocarbon (PAH), has shown serious toxicity towards a variety of plankton under environmentally stimulated. In the presence of natural or simulated sunlight, anthracene was severely toxic, at concentrations within aqueous solubility limits, to freshwater zooplankton. Experiments show that it takes less than 15%–50% immobilization of *Daphnia pulex*, a plankton species, at 1.2 anthracene/liter under natural sunlight.

6.3.4.4 Alkanolamines

Use of alkanol amines such as diethanolamine (DEA) in the removal of $CO_{(2)}$ from natural gas and for $CO_{(2)}$ capture following fossil fuel combustion. Apart from its variety of uses, little is known of its ecotoxicological impact (Sims & Quayle, 1998).

Research conducted to assess the impact of alkanol amines in the marine environment, an important species that exists in the North Atlantic, the planktonic copepod *Calanus finmarchicus*, shows the molecular and sub-lethal effects due to exposure to DEA in the marine environment (Siddons et al., 2022). Using suppression subtractive hybridization (SSH) and high-resolution magic angle spinning nuclear magnetic resonance (HR-MAS NMR), respectively, the alterations that occurred by introducing DEA in transcriptome and metabolism were estimated. Alterations in the transcription of genes involved in lipid metabolism, antioxidant systems, metal binding, and amino acid and protein catabolism were noted. These effects were accompanied by altered expression of fatty acid derivatives and amino acids such as methionine, glutamine, threonine, alanine, leucine, arginine, and choline. Relationships were also observed between DEA exposure and the transcription of genes associated with protein catabolism (ubiquitin-specific protease-7) and defense against oxidative stress (glutathione synthase and Cu/Zn-superoxide dismutase). Similar transcription patterns were observed for several different genes following exposure to DEA in *Calanus finmarchicus*, which indicates analogous mechanisms of toxicity (Morel, 2015).

6.3.4.5 Triclosan (TCS)

Triclosan (TCS), used as an antimicrobial agent in soaps and a variety of personal care and cosmetic products, halters the growth of bacteria by acting as a fatty acid biosynthesis inhibitor, hence, interrupting the TCA cycle and causing protein aggregation. It has been analyzed that the largely dangerous amounts of this chemical found in lakes have drastically reduced the lifespan of many plankton species in that habitat.

6.3.5 Why and How Should the Plankton Community Be Saved?

The earth is covered with 71% of the ocean and the oceans are mostly covered by phytoplankton. Therefore, phytoplankton is responsible for cycling most of the earth's carbon dioxide between the ocean and the atmosphere, thereby contributing to 50% of the oxygen that we breathe (Sugie et al., 2020). The phytoplankton, as their name suggests, use photosynthesis for survival. They are at the bottom of the food chain and are called the "primary producers". They act as food for the zooplankton, the primary consumers of the food web, which are in turn being eaten by small predators of the sea, the ones that are devoured by the larger animals. This way, the phytoplankton holds the food web upright, the absence of which would lead to its collapse. Plankton communities serve as good ecological indicators, as they are very sensitive to nutrient input in the water bodies (Case et al., 2008). Therefore, helping us supervise the ecological system quality and take immediate actions, for example, to prevent algal blooms, toxic contamination from undisclosed sources, etc. This is due to the property of plankton to react to even the slightest variations in their environment (Liu et al., 2021). Plankton communities are affected by various abiotic factors, such as light availability, temperature, salinity, heavy metals, pollutants, pH, and nutrient concentration, and biotic factors, such as predators, and parasites (Balushkina et al., 2009). The responses to variations in these factors are being studied through the ecological data collected to help us speculate where the plankton community varies and where there are harmful plankton species that grow rapidly due to the excessive nutrients in the water. The plankton community also has potential in both industrial and biotechnological aspects which enables them to function as main ingredients in making commercial produce. From an industrial perspective, phytoplankton and zooplankton species also serve as feedstock, wastewater treaters, or for the production of high-value compounds; and commercial products, such as food and feed supplements, pharmacological compounds, lipids, enzymes, biomass, polymers, toxins, and pigments. Intake of phytoplankton helps in muscle strengthening and also increases the metabolism of our body (Sommer et al., 2012). They also have amazing antioxidant properties that help in increasing the capacity of our cells (Hinkel, 2017).

Plankton is used as an ingredient in skincare products. They are composed of vitamin C which helps in collagen boosting and reducing hyperpigmentation. Since plankton resides mostly in water, they avoid UV radiation by going deep into water where UV radiation is less. Since plankton has such a unique and astounding ability, these are used in sunscreen which protects the skin from UV and other harmful radiation. They also have antibacterial properties which prevent the entry of toxins into the skin.

Potential measures should be taken by governments by improving waste management facilities to reduce plastic waste from entering the marine ecosystem and doing more research on how plastic waste affects the oceans. These solutions should be ingrained in government policies to ensure that efforts are effectively implemented and carried out, including serious measures on those corporations and individuals who breach these policies.

6.4 CONCLUSION

We have discussed the characteristics of the plankton community and their types. We have attained more knowledge on how they hold up our ecosystem and how we will be affected if the plankton community is distressed. The effects of waste from the pharma and cosmetic industries on these organisms have been thoroughly examined. The consequences of letting out plastics, microplastics, and organic and inorganic wastes have been proven to fatally affect these organisms. Apart from these, the adverse effects of cosmetic wastes such as parabens, alkanol amines, triclosan, etc., on the plankton community were reviewed. The reason why and how such pollutive activities should be prevented have been elucidated in this book chapter. By standing up for the smallest organisms in the ocean, we will be saving not only the animals in that habitat but also humans ourselves. When we reap so many benefits from the huge plankton communities, we must also concern ourselves with their nourishment and protection. If we wait to take a step until we face their extinction, humans will be in grave danger. So, remember to give thanks to the plankton that made it possible the next time you breathe deeply or savor some fresh seafood.

REFERENCES

Alfonina, Ekaterina Yu., and Natalya A. Tashlykova, 2018. Plankton community and the relationship with the environment in saline lakes of Onon-Torey plain, Northeastern Mongolia. https://www.sciencedirect.com/science/article/pii/S1319562X17300037

Ajani, Penelope A., Claire H. Davies, Ruth S. Eriksen, and Anthony J. Richardson, 2020. Global warming impacts micro-phytoplankton at a long-term pacific ocean coastal station. https://www.frontiersin.org/articles/10.3389/fmars.2020.576011/full

Balushkina, E. V., S. M. Golubkov, M. S. Golubkov, L. F. Litvinchuk, N. V. Shadrin, 2009. Effect of abiotic and biotic factors on the structural and functional organization of the saline lake ecosystems in Crimea. https://europepmc.org/article/med/20063772

Barton, A. D., A. Pershing, E. Litchman, N. R. Record, K. F. Edwards, Z. V. Finkel, T. Kiørboe, and B. Ward. 2013. The biogeography of marine plankton traits. *Ecology Letters*, 16, 522–534.

Beiras, Ricardo, and Tânia Tato, 2019. Microplastics do not increase toxicity of a hydrophobic organic chemical to marine plankton. https://www.sciencedirect.com/science/article/abs/pii/S0025326X18308063

Berthold, Maximilian, and Douglas A Campbell, 2021. Restoration, conservation and phytoplankton hysteresis. https://academic.oup.com/conphys/article/9/1/coab062/6350441

Borics, G., V. B-Béres, and I. Bácsi, 2020. Trait convergence and trait divergence in lake phytoplankton reflect community assembly rules. *Scientific Reports*, 10, 19599.

Bottrell, Zara L. R., Nicola Beaumont, Tarquin Dorrington, Michael Steinke, Richard C. Thompson, and Penelope K. Lindeque, 2019. Bioavailability and effects of microplastics on marine zooplankton: A review. https://www.sciencedirect.com/science/article/pii/S0269749118333190

Case, Maristela, Enide Eskinazi Leça, Sigrid Neumann Leitão, Eneida Eskinazi Sant'Anna, Ralf Schwamborn, and Antônio Travassos de Moraes Junior, 2008. Plankton community as an indicator of water quality in tropical shrimp culture ponds. https://www.sciencedirect.com/science/article/abs/pii/S0025326X08000805

Chia, Mathias Ahii, Adriana SturionLorenzi, Ilu Ameh, Suleiman Dauda, Micheline Kézia Cordeiro-Araújo, Jerry Tersoo Agee, Ibrahim Yusuf Okpanachi, and Abosede Taofikat Adesalu, 2021. Susceptibility of phytoplankton to the increasing presence of active pharmaceutical ingredients (APIs) in the aquatic environment: A review. https://www.sciencedirect.com/science/article/abs/pii/S0166445X21000680

Chiba, Sanae, Sonia Batten, Corinne S Martin, Sarah Ivory, Patricia Miloslavich, and Lauren V. Weatherdon, 2018. Zooplankton monitoring to contribute towards addressing global biodiversity conservation challenges. https://academic.oup.com/plankt/article/40/5/509/5079336

Chiba, S., S. D. Batten, T. Yoshiki, Y. Sasaki, K. Sasaoka, H. Sugisaki, and T. Ichikawa, 2015. Temperature and zooplankton size structure: Climate control and basin-scale comparison in the North Pacific. http://doi.org/10.1002/ece3.1408.

Costello, M. J., M. Coll, R. Danovaro, P. Halpin, H. Ojaveer, and P. Miloslavich, 2010. A census of marine biodiversity knowledge, resources, and future challenges. http://doi.org/10.1371/journal.pone.0012110.

Culliney, Kacey, 2020. Beauty and its wastes: Brands need to 'add value' to justify extra efforts and costs. https://www.cosmeticsdesign-europe.com/Article/2020/10/19/Waste-management-in-beauty-can-be-improved-if-brands-add-value-and-step-up-communication-says-Certified-Sustainable

Cullor, Ravyn, 2022. More plankton, more components: Improving marine toxicity testing of sunscreens. https://www.cosmeticsdesign.com/Article/2022/02/08/Study-on-testing-marine-toxicity-of-sunscreens-on-plankton

Desforges, J. P., C. Sonne, M. Levin, U. Siebert, S. De Guise, and R. Dietz, 2015. Immunotoxic effects of environmental pollutants in marine mammals. *Environment International*, 86, 126–139.

Dickinson, Kate, 2021. The use of cosmetics has become an indelible part of our everyday routines, and so too has the waste that comes with it. Just what is the extent of the beauty industry's throwaway culture, and can it be changed? https://resource.co/article/cosmetics-beautiful-waste-13018

Duarte, Bernardo, Carla Gameiro, Ana Rita Matos, Andreia Figueiredo, Marta Sousa Silva, Carlos Cordeiro, Isabel Caçador, Patrick Reis-Santos, Vanessa Fonseca, and Maria Teresa Cabrita. 2021. https://www.sciencedirect.com/science/article/abs/pii/S0045653521003295\

Florescu, Larisa I., Mirela M. Moldoveanu, Rodica D. Catană, Ioan Pacesila, Alina Dumitrache, Athanasios A. Gavrilidis, and Cristian I. Iojă, 2022. Assessing the effects of phytoplankton structure on zooplankton communities in different types of urban lakes. https://www.mdpi.com/1424-2818/14/3/231

Hinkel, Lauren, 2017. Plankton can save the ocean. But who will save the plankton? https://darwinproject.mit.edu/plankton-can-save-the-ocean-but-who-will-save-the-plankton/

Hitchcock, James N., 2022. Microplastics can alter phytoplankton community composition. https://www.sciencedirect.com/science/article/abs/pii/S0048969722001644

Istu Septania, 2021. *Saving the ocean's invisible forests.* https://earth.org/phytoplankton-saving-the-oceans-invisible-forests/

Jernberg, S., M. Lehtiniemi, and L. Uusitalo, 2017. Evaluating zooplankton indicators using signal detection theory. *Ecological Indicators.* http://doi.org/10.1016/j.ecolind.2017.01.038.

Jo, Naeun, Hyoung Sul La, Jeong-Hoon Kim, Kwanwoo Kim, Bo Kyung Kim, Myung Joon Kim, Wuju Son, and Sang Heon Lee, 2021. Different biochemical compositions of particulate organic matter driven by major phytoplankton communities in the northwestern Ross Sea. https://www.frontiersin.org/articles/10.3389/fmicb.2021.623600/full

Johns, D. G., M. Edwards, and S. D. Batten, 2001. Arctic boreal plankton species in the Northwest Atlantic. *Canadian Journal of Fisheries and Aquatic Sciences*, 58, 2121–2124. http://doi.org/10.1139/cjfas-58-11-2121.

Kim, Dokyun, Ha-Eun Cho, Eun-Ji Won, Hye-Jin Kim, Sunggyu Lee, Kwang-Guk An, Hyo-Bang Moon, Kyung-Hoon Shin, 2022. Environmental fate and trophic transfer of synthetic musk compounds and siloxanes in Geum River, Korea: Compound-specific nitrogen isotope analysis of amino acids for accurate trophic position estimation. https://www.sciencedirect.com/science/article/pii/S0160412022000496

Kiørboe, T., and A. G. Hirst. 2014. Shifts in mass scaling of respiration, feeding, and growth rates across lifeform transitions in marine pelagic organisms. *American Naturalist*, 183, E118–E130.

Liu, Qi, Yulu Tian, Yang Liu, Ming Yu, Zhaojiang Hou, Kejian He, Hui Xu, Baoshan Cui, and Yuan Jiang, 2021. Relationship between dissolved organic matter and phytoplankton community dynamics in a human-impacted subtropical river. https://www.sciencedirect.com/science/article/abs/pii/S095965262035188X

Mackas, D. L., S. Batten, and M. Trudel, 2007. Effects on zooplankton of a warmer ocean: Recent evidence from the Northeast Pacific. *Progress in Oceanography.* http://doi.org/10.1016/j.pocean.2007.08.010.

Mangolte, Inès, Marina Lévy, Stephanie Dutkiewicz, Sophie Clayton, and Oliver Jahn, 2022. Plankton community response to fronts: Winners and losers. https://academic.oup.com/plankt/article-abstract/44/2/241/6535703?redirectedFrom=fulltext

Möhlenkamp, Paula, Autun Purser, and Laurenz Thomsen, 2018, September 7. Plastic Microbeads from cosmetic products: An experimental study of their hydrodynamic behavior, vertical transport and resuspension in phytoplankton and sediment aggregates. *Plastic Microbeads from Cosmetic Products: An Experimental Study.* https://online.ucpress.edu/elementa/article/doi/10.1525/elementa.317/112836/Plastic-microbeads-from-cosmetic-products-an

Morel, François, 2015. Effects of ocean acidification on marine phytoplankton. https://cmi.princeton.edu/annual-meetings/annual-reports/year-2015/effects-of-ocean-acidification-on-marine-phytoplankton/

Murphy, Grace E. P., Tamara N. Romanuk, and Boris Worm, 2020. Cascading effects of climate change on plankton community structure. https://onlinelibrary.wiley.com/doi/full/10.1002/ece3.6055

Murrell, Michael C., and Emile M. Lores, 2004. Phytoplankton and zooplankton seasonal dynamics in a subtropical estuary: Importance of cyanobacteria. https://academic.oup.com/plankt/article/26/3/371/1442607

Sally, Gaw, Kevin V. Thomas, and Thomas H. Hutchinson, 2019. Sources, impacts and trends of pharmaceuticals in the marine and coastal environment. https://www.ncbi.nlm.nih.gov/pmc/articles/PMC4213585/

Sharp, Matthew, Kazim Sahin, Matthew Stefan, Cemal Orhan, Raad Gheith, Dallen Reber, Nurhan Sahin, Mehmet Tuzcu, Ryan Lowery, Shane Durkee, and Jacob Wilson, 2020. Phytoplankton supplementation lowers muscle damage and sustains performance across repeated exercise bouts in humans and improves antioxidant capacity in a mechanistic animal. https://www.ncbi.nlm.nih.gov/pmc/articles/PMC7400322/

Siddons, Joseph T., Andrew J. Irwin, and Zoe V. Finkel, 2022. Graphical analysis of a marine plankton community reveals spatial, temporal, and niche structure of sub-communities. https://www.frontiersin.org/articles/10.3389/fmars.2022.943540/full

Sims, D. W., and V. A. Quayle, 1998. Selective foraging behavior of basking sharks on zooplankton in a small-scale front. http://doi.org/10.1038/30959.

Sommer, Ulrich, 2002. Population dynamics of phytoplankton. http://eolss.net/Sample-Chapters/C09/E2-27-03-02.pdf

Sommer, Ulrich, Rita Adrian, Lisette De Senerpont Domis, James J. Elser, Ursula Gaedke, Bas Ibelings, Erik Jeppesen, Miquel Lürling, Juan Carlos Molinero, Wolf M. Mooij, Ellen van Donk, and Monika Winder, 2012. Beyond the Plankton Ecology Group (PEG) model: Mechanisms driving plankton succession. https://www.annualreviews.org/doi/abs/10.1146/annurev-ecolsys-110411-160251

Spilling, Kristian, Letizia Tedesco, Riina Klais, and Kalle Olli, 2019. Editorial: Changing plankton communities: Causes, effects and consequences. https://www.frontiersin.org/articles/10.3389/fmars.2019.00272/full#:~:text=The%20plankton%20community%20makes%20up,Falkowski%20and%20Raven%2C%202013).

Stone, Kathlyn, Gurdeep S. Sareen, and Angela Underwood, 2023. What is an active pharmaceutical ingredient? https://www.verywellhealth.com/api-active-pharmaceutical-ingredient-2663020

Sugie, Koji, Amane Fujiwara, Shigeto Nishino, Sohiko Kameyama, and Naomi Harada, 2020. Impacts of temperature, CO2, and salinity on phytoplankton community composition in the Western Arctic Ocean. https://www.frontiersin.org/articles/10.3389/fmars.2019.00821/full

Tovar-Sánchez, Antonio, David Sánchez-Quiles, Gotzon Basterretxea, Juan L. Benedé, Alberto Chisvert, Amparo Salvador, Ignacio Moreno-Garrido, and Julián Blasco, 2013. Sunscreen products as emerging pollutants to coastal waters. https://www.ncbi.nlm.nih.gov/pmc/articles/PMC3673939/

Tweddle, Jacqueline F., Matthew Gubbins, and Beth E. Scott, 2018. Should phytoplankton be a key consideration for marine management? https://www.sciencedirect.com/science/article/pii/S0308597X17300854

Weber, Cornelius I., 1968. The preservation of phytoplankton grab samples. https://www.jstor.org/stable/3224339

Yakubu, A. F., T. E. Adams, E. D. Olaji, E. E. Adebote, and O. R. Okabe, 2018. Effect of inorganic fertilization in the zooplankton production in freshwater pond. https://ijeab.com/upload_document/issue_files/22-IJEAB-OCT-2018-31-EffectofInorganic.pdf

7 Zooplankton and Phytoplankton Community

Influence of Personal Care Products, Legislative Status, and Possible Remediation

Osikemekha Anthony Anani,
Kenneth Kennedy Adama,
Paul Atagamen Aidonojie, and
Okagbare Aregbor

7.1 INTRODUCTION

An assorted collection of compounds such as sunscreens, fragrances, toothpaste, body lotions, soaps, etc., used for body grooming and cleanness is called PCPs (personal care products) (Brausch and Rand, 2011; Sotão Neto et al., 2020). The 1° of PCPs classes includes methylbenzylidene camphor, parabens, insect repellants, musks, and triclosan (Brausch and Rand, 2011). The products of PCPs are mainly utilized externally on the human body which does not need to undergo some biological and physiological alterations.

McKnight et al. (2015) reported that hundreds of chemicals are utilized daily around the world, and their end products are usually leached into aquatic environments and accumulate into the tissues of living organisms. Hence, enormous amounts of PCPs are leached into the environment during human use and are unaltered via consistence and persistent use (Brausch and Rand, 2011). Lots of PCPs are used by humans and recent studies showed that being bioactive and persistent in the environment, they may bioaccumulate and show probable risk to the environment and health risks via the structure of the food chain (Anani et al., 2022, 2023).

Studies have shown that triclocarban and triclosan which are biphenyl ethers mainly found in plastics, toothpaste, skin creams, deodorants, and soaps are the most ten tops frequent chemicals found in aquatic environments via wastewater input of about 10 lg L^{-1} in concentration (Brausch and Rand, 2011). Fragrances like nitro musks (that have faded out) and artificial musks used in the array of products like

 DOI: 10.1201/9781003362975-7

detergents, soaps, and deodorants have been shown to have probable toxicity to aquatic biota (Daughton and Ternes, 1999). Insect repellants like the N, N-diethyl-m-toluamide have also been found in the surface water in several regions of the world (Glassmeyer et al., 2005; Costanzo et al., 2007; Brausch and Rand, 2011). USEPA (1998), Costanzo et al. (2007), and Brausch and Rand (2011) reported about 1.8 million kg of N, N-diethyl-m-toluamide in the surface water and stated the persistent nature of the pollutants in the aquatic environment and in the tissues of the biotas. Parabens which have been used as food, pharmaceuticals, and cosmetic preservatives for many decades have also been found on surface water with a range of concentration of 15 to 400 ng L^{-1} (Brausch and Rand, 2011). UV filters that are utilized in cosmetics and sunscreen materials for body grooming to protect the body from radiation (ultraviolet) have been also found to be noxious to aquatic organisms when leached into the environment. Daughton and Ternes (1999) have reported the incidences of accumulation of UV filters in the muscles of aquatic organisms like fish at a level that is almost the same as DDT and PCBs.

The consortia of the phytoplankton and the zooplankton in the aquatic environment are faced with many intrusions of chemicals that may affect their breeding, sexual behaviors, and ecological niche. Emergent pollutants like the chemicals found in PCPs have the potential to cause serious harm to the plankton communities in the freshwater or marine environment. So there is a need to evaluate the possible effects of these chemicals. Hence, this chapter reviews the personal care products (PCPs), cosmetics, as persisting pollutants in zooplankton and phytoplankton communities.

7.2 CURRENT CHALLENGES OF PCPS IN THE AQUATIC ENVIRONMENT

Ordinarily, personal care products (PCPs) are pollutants referred to as organic micropollutants belonging to a group of compounds that find biological applications in medicine, veterinary medicine, and the maintenance of daily human hygiene. The presence of PCPs even in trace quantities or concentrations in the environment has negative effects on the abiotic and biotic environment. The presence of PCPs in the environment is caused largely by the improper production, usage, and disposal of cosmetics and other medical drugs. PCPs can be introduced into the environment such as wastewater, water, and soil due largely to human activities.

The PCPs are a group of materials that includes food supplements, nutrients, various types of cosmetics, and their additives such as antiseptics and antibacterial properties. They shampoo, UV blockers, toilet water, and antiseptics. Others in this group include disinfectants (antibacterial and antiviral), which have increasingly been used since 2020 due largely to the outbreak of the dreaded COVID-19 (Anand et al., 2022; Adhikari et al., 2022; Wydro et al., 2024). PCPs show great varieties of physicochemical properties that include a complex chemical structure and stability making their evaluation, analysis, and removal difficult (Wydro et al., 2024). Additionally, due to the growing number of personal hygiene products introduced into the market globally, the number of studies on the removal of PCPs in recent times from water and other aquatic environment and their impact on the environment is minimal.

Currently, PCPs concerning cosmetics have been identified as a persistent pollutant in most aquatic communities which has greatly attracted serious concerns worldwide due to the detrimental effects and presence of active and toxic compounds in its constituent's composition, thereby posing hazards to the aquatic environment, human beings, and other wildlife (Miller et al., 2018; Duan et al., 2022). Guitart and Readman (2010) and Fahlman et al. (2021) reported incidents level of about 70% due to the presence of these chemicals in the PCPs over the last two decades. Thus, the prevalence of cosmetics as personal care products in aquatic environments has been observed to compromise the delivery of some of the SDG goals which are clean water and sanitation (Goal 6), responsible consumption and production (Goal 12), life below water (Goal 14), and life on land (Goal 15) (Kock et al., 2023).

Personal care products (PCPs) have found utilization in several applications from human activities to nonhuman activities over time and space. The essential utilization and beneficial uses have continued to attract attention as much as the inherent challenges associated with its use in biotic and abiotic situations or environments. Personal care products are a widely used group of substances that have inherent health concerns which when released into the aquatic environment cause substantial detrimental effects on the ecosystem. Molins-Delgado et al. (2014) reported on the occurrence and impact of ingredients in personal care products on the aquatic environment taking cognizance of the methodologies of analysis, prevalence data utilized, possible elimination processes, inherent threats to the aquatic ecosystem, and the effects on biota as well as possible legislations. All activities in nature require water as an available resource and thus covers most aspects of the earth's surface. Due to population increases globally, there has been an intense use of water resources for human activities, thereby generating huge environmental pollutants in the aquatic water body over time. Different studies reveal that the continuous application on the skin of PCPs or the intake of contaminated PCPs food may greatly affect human beings. To ensure the protection of the aquatic ecosystem, there have been concerted initiatives geared towards the development of novel monitoring and governmental approaches worldwide (Brausch and Raud, 2011; Molins-Delgado et al., 2014).

Although considerable research has been conducted on the occurrence and effects of human use of pharmaceuticals in the aquatic environment; however, there has been relatively little research on personal care products in aquatic ecosystems, even though their occurrence is more prevalent and in higher concentrations than pharmaceuticals in aquatic water bodies. Personal care products are continually released into the aquatic environment and are biologically active and persistent. Much data on the acute and chronic toxicity nature of cosmetics products are available for personal care products, and this highlights areas of concern according to Brausch and Raud (2011). According to Kock et al. (2023), the presence of personal care products in aquatic environments presents risks to inhabiting plant and animal species in such ecosystems. The persistence of PCPs in the aquatic environment is an important issue that has attracted global concerns from the scientific community over time (Gomaa et al., 2021; Mojiri et al., 2021; Ngqwala and Muchesa, 2020; Xin et al., 2020; Wang et al., 2017b). Environmental factors such as chemical and biological properties contained in the PCPs have been identified as some of the pollution indicators in many aquatic environments. Processes in the environment such as absorption, photocatalytic decomposition, nutrient depletion distribution, and operational metabolism of living organisms within the biota have been known to influence the biochemical reaction within the aquatic environment (Xin et al., 2020). The quantities of PCPs present in the environment could be as little to moderate in concentration. These still pose a great threat to the aquatic system, as they tend to bioaccumulate in the water body due largely to their inability to degrade substantially and, more importantly, consistent daily use and continuous release (Ngqwala and Muchesa, 2020).

Different categorizations of PCPs have evolved. This PCP categorization includes cosmetics and personal hygiene products, synthetic musk fragrances such as nitro polycyclic musks, insect repellents, UV blockers such as methylbenzylidene camphor, and preservatives which include phenols and p-hydroxybenzoic acid also called parabens. Parabens are actively used in cosmetics and include shampoos and balms. They are antimicrobial preservatives that possess excellent water solubility and stability. Their presence in various elements of the environment is disturbing because studies have shown that parabens may cause breast cancer in women and sperm dysfunction in men (Wydro et al., 2024). Parabens find applications extensively as preservatives in cosmetics, foodstuffs, and pharmaceuticals.

7.3 ZOOPLANKTON AND PHYTOPLANKTON COMMUNITIES IN FRESHWATER AND MARINE ENVIRONMENTS AND HOW THEY THRIVE AGAINST POLLUTANTS

Wei et al. (2022) studied the various responses zooplankton and phytoplankton communities react to extant altering of coastal environments. Organisms in the coastal ecosystem are faced with many physical, chemical, and biological factors such as eutrophication, warming, and acidification which have significantly reduced the communities of zooplankton and phytoplankton. The incidences of pH and temperature have been found to have great effects on the decline of the biota's community as well. Wei et al. (2022) also stated that As, Zn, and Hg have been shown to have diverse implications on the communities. The recommended that if this is not curtailed via environmental remediation, the aquatic biotas may reduce in the future under the influence of these factors.

Duarte et al. (2023) examined the effects of emerging contaminants like diclofenac and sulfamethoxazole on the aquatic ecosystem and to the phytoplankton community therein. Anthropogenic activities have been shown to cause the input of chemicals like pesticides, herbicides, emergent wastes, and sewage to leach into groundwater which has resulted in pollution. The presence of wastes from emergent inputs can change the water status of the ecosystem and also affect the composition of the biota therein. By employing an experimental technique, Duarte et al. (2023) did a laboratory assessment on the community of phytoplankton for 15 (fifteen) days using the following concentrations: 0.1, 0.5, and 1.0 mg L^{-1}. Six groups of green algae and diatoms were identified with varied density and diversity of organisms when diclofenac and sulfamethoxazole drugs were administered. The findings from the results showed that there was a disparity in the cyanobacteria abundance when they were treated with diclofenac. Meanwhile, when desmids were treated with sulfamethoxazole, similar effects were observed. This confirmed that the attendance of pharmaceuticals in aquatic environments can affect the community of phytoplankton, especially their abundance and diversity. Duarte et al. (2023) recommended the consistent evaluation of the aquatic ecosystem for pharmaceuticals and their possible influence on the biota therein to establish a better framework for the management of water bodies from such emergent pollutants.

Ianora et al. (2011) evaluated the effects of emergent chemicals or pollutants in the marine environment on plankton and the chemical cues they use to deal with the immediate environment. The authors stated that the environment of saltwater consists of chemicals of bioactive origin that affect the survival of phytoplankton (dinoflagellates and prymnesiophytes) and zooplanktons (dinoflagellates, diatoms, and copepods) in the marine ecosystem. However, most of the planktons can defend themselves against pollutants by developing a barrier system—using cell-to-cell gesturing as a buffer medium as found in diatoms. At times, this chemical cue cannot aid in the total protection of the plankton species because of the specifics of the bioactive ingredients produced by the innate physiological cell. Understanding this physiological process, one will tend to understand the physiological and chemical processes that enable plankton to thrive successfully in their immediate environment.

7.4 LEGISLATIVE STATUS OF PCPs

In accordance to the extant European Union legislation on PCPs, "any material intended to be used on contact surfaces like the oral cavity, genitalia, lips, nails, hair, mucous membrane, and the epidermis with the aim of body grooming, is termed personal care products" (Anonymous, 1976, 1993; Morganti and Paglialunga, 2008).

For almost 30 decades, 1994 precisely, some consumers of PCPs filed three legal suits on some companies that produce hair creams because of the health effects—alopecia it caused during their utilization (Martin, 2020; Rizzi, 2020; Flanagan et al., 2021). They were given the sum of $4.5 million for settlement due to the evidence of the presence of formaldehyde in the hair cream. Several cases of scalp irritation, rashes, pruritus, and loss of hair were also established by 21,000 protesters to the United State FDA (Food and Drug Administration) with about $26 million awarded to the consumers of hair care products called WEN (Rizzi, 2020; Flanagan et al., 2021).

Even though consumers of PCPs have assumed the safety of the ingredients used in most PCPs, there is the possibility of adverse health effects when the chemicals used in their preparation are not well monitored by regulating bodies like the FDA. Flanagan et al. (2021) reported that most PCPs do not undergo proper scrutiny because of the limitations set by the FDA. There is no clear define approval for the ingredients and products used in their preparation but rather the color flavors.

More so, the manufacturers of the PCPs are not allowed to do specific adverse event and safety tests to ascertain the potential harms the products portend. The FDA can only control personal care products under the Federal Cosmetic, Drug, and Food Act of 1938 as well as the Labelling, Packaging, and the Fair Act of 1967 which permit both manufacturers to utilize ingredients that are not adulterated or misbranded products.

Currently, the FDA explained that any products that is decomposed, decaying, and filthy is classified under adulterated products. In light of this, the previous laws that is now outdated (1938 and 1967 acts) need to be reviewed to improve the customers safety and their health too. The PCPs and Cosmetic Act of 2019 which stopped superfluous ingredients, allergic, and noxious ingredients, like the paraphenylenediamine, formaldehyde releasers, and the formaldehyde, were advocated by dermatologists to safeguard the health of their patients and the end consumers of PCPs products that are toxic (Flanagan et al., 2021). The act is now been used to regulate adulterated chemical and their use in PCPs today.

Under this act, it is required for manufacturers to submit SI (safety information) on any PCPs relation ingredients 15 days to the FDA to ascertain if any adverse events (ADs) or serious health effects would arise on the use of the PCPs and make it available to the public domain. If there is a possibility of any ADs, the bill/act also gives the FDA the prerogative to recall any products if found unsafe for public consumption (Flanagan et al., 2021).

7.5 POTENTIAL WAYS TO REMEDIATE POLLUTED ENVIRONMENT FROM PCPS THAT ARE NOXIOUS

The potential ways to remediate polluted environments from PCPs that are noxious could be by the use of conventional separation processes or advanced methods. In general, techniques for the elimination of PCPs in aquatic environment include slow filtration with sand filters, ozonization, techniques based on the advanced oxidation processes (AOPs) and electrochemical oxidation, adsorption on granular activated carbon (GAC), and membrane techniques, in particular, nanofiltration and reverse osmosis. In the case of wastewater treatment, conventional biological processes are often insufficient. Therefore, membrane bioreactors are recommended

Conventional polluted water treatment techniques are aimed at removing pathogens, reducing color, turbidity, and controlling odor and taste. However, these methods are not efficient in completely eliminating organic micropollutants (PCPs). Therefore, it is necessary to introduce advanced methods. Among the currently used advanced methods are adsorption methods, AOPs, membrane separation techniques, and combined technologies (Figure 7.1).

7.5.1 ELECTROCOAGULATION

One of the methods being used to remove antibiotics and PCPs from water is electrocoagulation. The main processes used in electrocoagulation are charge adsorption and neutralization. The main reactions occurring in the electrocoagulation process are the dissolution of metal in the anodic part and the production of hydroxyl anion in the cathodic part.

7.5.2 ADVANCED OXIDATION PROCESSES

Nowadays, commonly used methods that allow for the elimination of micropollutants from wastewater and water are the advanced oxidation processes (AOPs).

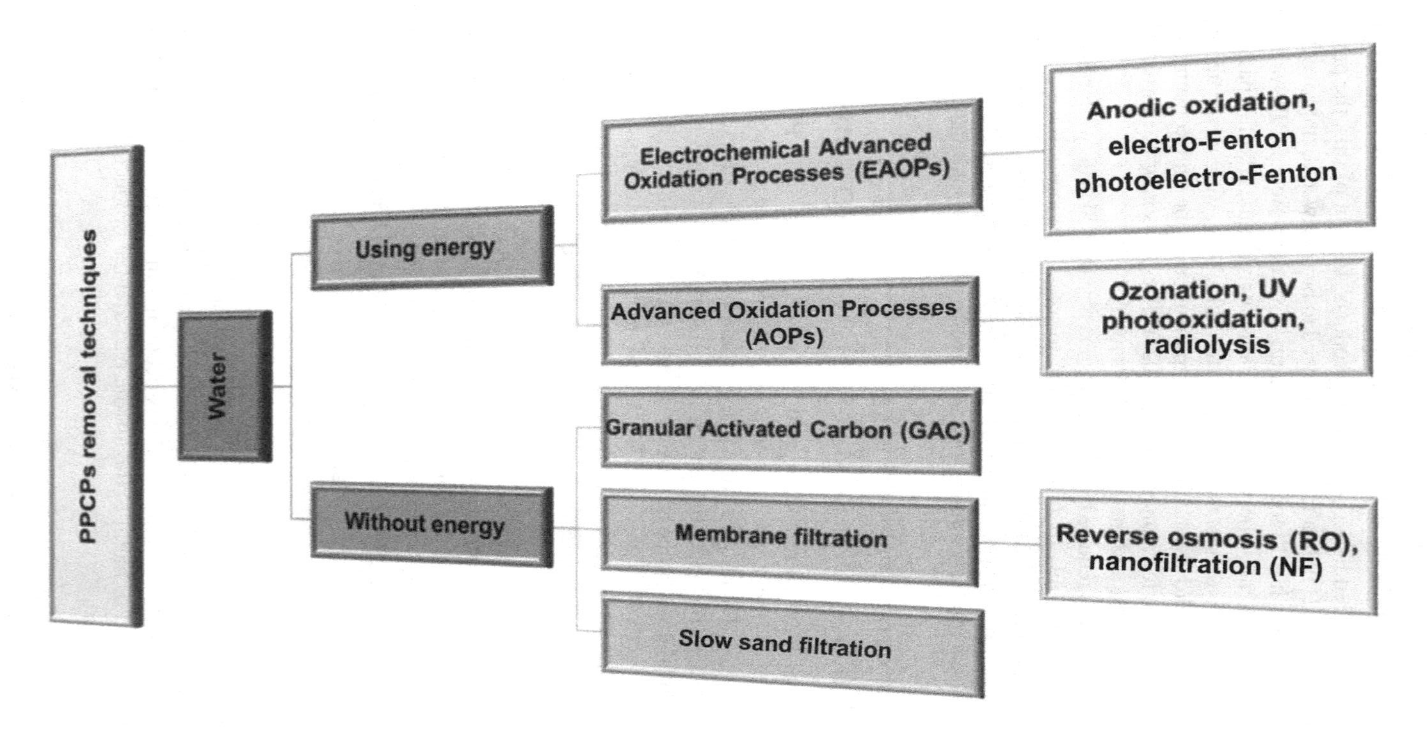

FIGURE 7.1 Methods of removing PCPs from polluted aquatic water body.

Source: Wydro et al. (2024).

AOPs are defined as processes in which a strong oxidant, normally the hydroxyl radical, is produced and used, with the property of oxidizing virtually any organic compound to carbon dioxide, water, and inorganic compounds. In order to obtain the requisite radicals, the use of hydrogen peroxide, ozone, UV radiation, additives of catalysts (MnO2, Fe^{2+}, and TiO_2), and their combinations are employed. For enhanced utilization and beneficial results, it is recommended to use systems containing use or three components such as O_3/UV, O_3/H_2O_2, H_2O_2/UV, or O_3/H_2O_2/UV, respectively. Over time, it is becoming more apparent that technologies based on advanced oxidation have become more intensively developed and this has attracted a lot of scientific interest. This development emanated from the fact that the complete oxidation of the polluted PCPs environment has resulted in the absence of the harmful and noxious by-products. Amongst the several types of advanced oxidation technologies, the photocatalytic method of remediating polluted environment from PCPs happens to be the most promising. This is as a result of its high degree of degradation and ability to convert such pollutants to mineral substances. This technology employs the use of sunlight such that impurities resistant to decomposition like organic matter, inorganic matter, and other pollutants could be broken down and converted to simple compounds such as CO_2 and H_2O.

7.5.3 Advanced Electrochemical Oxidation Processes

Advanced electrochemical oxidation processes (EAOPs), such as anodic oxidation (AO), electro-Fenton (EF), and photoelectro-Fenton (PEF), can be employed to remediate polluted environment from PCPs that are noxious. The inherent properties of electrochemical oxidation, occasioned by its strong oxidation performance, environmental compatibility, and mild reaction conditions, provides the needed impetus for its use in remediating polluted environments from PCPs. However, the processes are noted to operate extensively on high energy utilization, and this has significantly hampered its commercial use. EAOPs technology also include membrane technologies. Electrodialysis is the simplest membrane method based on electrochemical technology that uses ion-exchange membranes.

7.5.4 Adsorption Processes

Another potential way to remediate polluted environment from PCBs is the use of various adsorption methods. These methods are effective in remediating organic micropollutants from polluted environments containing different PCPs. The use of various substances as adsorbents is highly recommended, such as activated carbon, chitosan, resins, zeolites, or waste-based adsorbents. Bioadsorbents from graphene-based nanoadsorbent materials, carbon nanotubes, and biochar have also found application for remediating polluted environments. In general, adsorption methods even with those using biochar have minimal efficiency, though they are of low cost and require low levels of technological advancement. Additionally, the adsorbent could be obtained from waste materials that invariably lowers the costs of production and enhances a sustainable and efficient waste management policy.

7.5.5 Membrane Filtration Processes

Membrane filtration processes could find use as potential ways to remediate polluted environments containing PCPs. Among the membrane filtration methods used to remediate polluted environments from micropollutants and PCPs are nanofitration (NO) and reverse osmosis. Microfiltration and ultrafiltration membranes in addition to biological treatment processes formed the basis in the development of membrane bioreactors. The limitation in the use of micro- and ultrafiltration membranes for the elimination of PCPs from polluted environments is the higher molecular weight of the membrane as compared to the molecular weight of most PCPs. However, owing to this enormous challenge, the current use of the membrane filtration processes encourages the use of modified or hybrid versions of membrane processes to meet current realities.

7.5.6 Microalgae and Phytoremediation Technology

Currently, one of the key biotechnological techniques employed to remediate polluted environment from PCPs is the use of algae. This is so because of the inherent properties of the system that include environmental responsiveness, ease of culture, and several biological assets. The technology also has applications in the removal of PCPs from polluted environment and forms one of the several potential ways to remediate contaminated aquatic environments.

The occurrence of cosmetics as pollutants in aquatic environment is particularly dangerous to the plant and animal life in aquatic environment. The earlier analysis of various remediation technologies allows us to conclude that it is possible to eliminate most PCPs using advanced remediation methods (Figure 7.2). Therefore, legislative issues are important and require quick solutions to affect these preventive and corrective approaches.

7.6 PROSPECTS ON PCPs REMEDIATION

In line with the need to remediate PCPs and related emerging pollutants in the environment, several authors have recommended some possible remediation if found in any environmental matrixes like water and soil. Adeleye et al. (2022), Ghosh et al. (2023), and Wydro et al. (2024) suggest the use of biological, chemical, and physical in the remediation of PCPs specifically found in wastewater, sludge, and sediments from pharmaceutical and veterinary wastes effluents.

However, because of the recalcitrant nature of most PCPs in sewage and water system, Wydro et al. (2024) recounted that there is a need to deploy AOPs (advanced oxidation processes) and GAC (granular activated carbon) techniques in the decontamination and removal of PCPs in this media because of their efficiencies and reliability. The use of a hybrid technique like microalgae, membrane bioreactors, and AOPs are also needed because of their efficiency to remove, primary, secondary, and tertiary PCPs micropollutants at every phase, thus making it environmentally friendly.

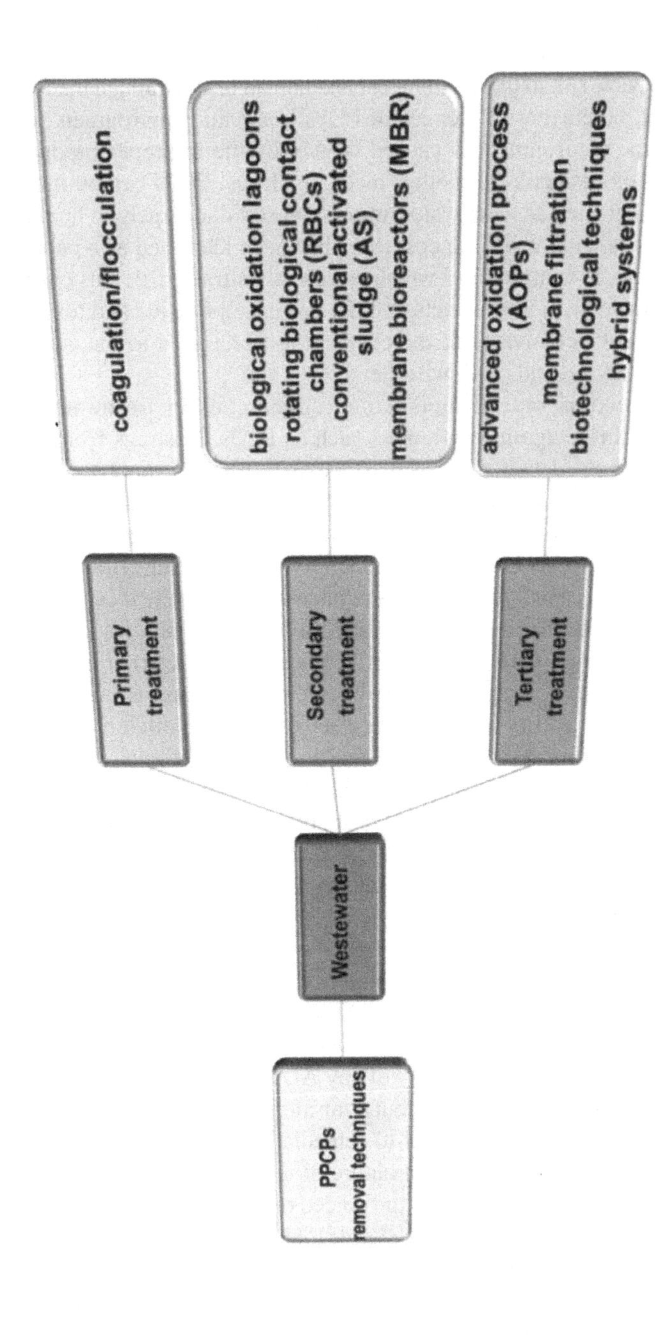

FIGURE 7.2 Methods of remediating polluted environment from PCPs.

Source: Wydro et al. (2024).

7.7 CONCLUSION

This chapter looks at PCPs as persistent potential pollutants in zooplankton and phytoplankton communities. Lots of PCPs are used by humans, and recent studies have shown that due to their bioactivity and persistence in the environment, they may bioaccumulate and show probable risk to the environment and human health through the food chain.

As regard the current challenges of PCPs in aquatic environment, the presence of PCPs in the environment is caused largely by the improper production, usage, and disposal of cosmetics and other medicals drugs. PCPs can be introduced into the environment such as wastewater, water, and soil due largely to human activities. Currently, PCPs with respect to cosmetics has been identified as a persistent pollutant in most aquatic communities which has greatly attracted serious concerns worldwide due to the detrimental effects and the presence of active and toxic compounds in its constituent's composition, thereby posing as hazards to the aquatic environment, human beings, and other wildlife.

The zooplankton and phytoplankton communities in freshwater and marine environments thrive against pollutants such as PCPs. Findings from the results of some studies showed that there was a disparity in the cyanobacteria abundance when they were treated with diclofenac. Meanwhile, when desmids were treated with sulfamethoxazole, similar effects were observed. This confirmed that the attendance of pharmaceuticals in aquatic environment can affect the community of phytoplankton, especially their abundance and diversity. It is recommended to consistently evaluate the aquatic ecosystem for pharmaceuticals and their possible influence to the biota therein. This will help establish a better framework for managing water bodies affected by such emergent pollutants. However, most of the planktons have the ability to defend themselves against pollutants by developing a barrier system—using cell-to cell gesturing as a buffer medium as found in diatoms. At times, this chemical cue cannot aid in total protection of the plankton species because of the specifics of the bioactive ingredients produced by the innate physiological cell. Understanding this physiological process, one will tend to understand the physiological and chemical processes that enable plankton to thrive successfully in their immediate environment.

Therefore, the legislative status of PCPs showed that under the 2019 Act on PCPs and Cosmetic, it is required for manufacturers to submit SI (safety information) on any PCP-related ingredients 15 days to the FDA to ascertain if any adverse events (ADs) or serious health effects would arise on the use of the PCPs and make it available to the public domain. If there is a possibility of any ADs, the bill/act also gives the FDA the prerogative to recall any products if found unsafe for public consumption.

Several methods have been used to remediate PCPs in wastewater and sludge. However, because of the recalcitrant nature of most PCPs in sewage and water systems, there is a need to deploy AOPs (advanced oxidation processes) and GAC (granular activated carbon) techniques in the decontamination and removal of PCPs in this media because of their efficiencies and reliability. The use of a hybrid technique like microalgae, membrane bioreactors, and AOPs are also needed because of their efficiency to remove primary, secondary, and tertiary PCPs micropollutants at every phase, thus making it environmentally friendly.

REFERENCES

Adeleye, A.S., Xue, J., Zhao, Y., Taylor, A.A., Zenobio, J.E., Sun, Y., Han, Z., Salawu, O.A., Zhu, Y. (2022). Abundance, fate, and effects of pharmaceuticals and personal care products in aquatic environments. *Journal of Hazardous Mater.* 424: 127284.

Adhikari, S., Kumar, R., Driver, E.M., Perleberg, T.D., Yanez, A., Johnson, B., Halden, R.U. (2022). Mass trends of parabens, triclocarban and triclosan in Arizona wastewater collected after the 2017 FDA ban on antimicrobials and during the COVID-19 Pandemic. *Waste Research.* 222: 118894.

Anand, U., Adelodun, B., Cabrevos, C., Kumar, P., Suresh, S., Dey, A., Ballesteros, F., Bontempi, E. (2022). Occurrence, transformation, bioaccumulation, risk and analysis of pharmaceutical and personal care products from wastewater: A review. *Environmental Chemistry Letters.* 20: 3883–3904.

Anani, O.A., Adetunji, C.O., Anani, G.A., Olomukoro, J.O., Imoobe, T.O.T., Enuneku, A.A., Tongo, I. (2022). Effect of meso-, micro-, and nano-plastic waste on the benthos. In: Shahnawaz, M., Sangale, M.K., Daochen, Z., Ade, A.B. (eds) *Impact of Plastic Waste on the Marine Biota.* Springer, Singapore. https://doi.org/10.1007/978-981-16-5403-9_12.

Anani, O.A., Shah, M.P., Aidonojie, P.A., Enuneku, A.A. (2023). Bio-Nano filtration as an abatement technique used in the management and treatment of impurities in industrial wastewater. In: Shah, M. P. (ed) *Bio-Nano Filtration in Industrial Effluent Treatment.* England & Wales, London, CRC-Taylor and Francis, 169–182. https://doi.org/10.1201/9781003165149-11.

Anonymous (1976). *Council Directive 76/768/EEC of 27 July 1976.* On the Approximation of the Laws of the Member States Relating to Cosmetic Products. O.J.E.C. n L262/169. 27.9.1976 (and following amendments).

Anonymous (1993). *Council Directive 93/35/EEC of 14 June 1993.* Amending for the Sixth Time Directive 76/68/EEC on the Approximation of the Laws of the Member States Relating to Cosmetic Products. O.J.E.C. n. L 151/32, 23.6.1993.

Brausch, J.M., Rand, G.M. (2011). A review of personal care products in the aquatic environment: Environmental concentrations and toxicity. *Chemosphere.* 82(11): 1518–1532.

Costanzo, S.D., Watkinson, A.J., Murby, E.J., Kolpin, D.W., Sandstrom, M.W. (2007). Is there a risk associated with the insect repellant DEET (N,N-diethyl-mtoluamide) commonly found in aquatic environments? *Science of a Total Environment.* 384: 214–220.

Daughton, C.G., Ternes, T.A. (1999). Pharmaceuticals and personal care products in the environment: Agents of subtle change? *Environmental Health Perspectives.* 107: 907–937.

Duan, W., Cui, H., Huang, X. (2022). Occurrence and ecotoxicity of sulfonamides in the aquatic environment: A review. *Science of a Total Environment.* 153178

Duarte, J.A.P., Ribeiro, A.K.N., de Carvalho, P., Bortolini, J.C., Ostroski, I.C. (2023). Emerging contaminants in the aquatic environment: Phytoplankton structure in the presence of sulfamethoxazole and diclofenac. *Environmental Science and Pollution Research International.* 30(16): 46604–46617. http://doi.org/10.1007/s11356-023-25589-2.

Fahlman, J., Hellstrom, G., Jonnson, M., Fick, J.B., Rosvall, M., Klaminder, J. (2021). Impacts of oxazepam on perch (percafluviatilis) behaviour: Fish familiarized to lake conditions do not show predicted anti-anxiety response. *Environmental Science Technology.* 55(6): 3624–3633.

Flanagan, K.E., Pathoulas, J.T., Walker, C.J., Wiss, I.P., Senna, M.M. (2021). Legislative update: Regulating ingredients in personal care products. *Health Policy and Practice.* The American Academy of Dermatology, Inc. https://doi.org/10.1016/j.jaad.2021.02.025.

Ghosh, S., Nandasana, M., Webster, T.J., Thongmee, S. (2023). Agrowaste-generated biochar for the sustainable remediation of refractory pollutants. *Frontier in Chemistry.* 16(11): 1266556. http://doi.org/10.3389/fchem.2023.1266556.

Glassmeyer, S.T., Furlong, E.T., Kolpin, D.W., Cahill, J.D., Zaugg, S.D., Werner, S.L., Meyer, M.T., Kryak, D.D. (2005). Transport of chemical and microbial compounds from known wastewater discharges: Potential for use as indicators of human fecal contamination. *Environment Science Technology*. 39: 5157–5169.

Gomaa, M., Zien-Elabdeen, A., Hifriey, A.F., Adam, M.S. (2021). Environmental risk analysis of pharmaceuticals on freshwater phytoplankton assemblage: Effects on alpha, beta, and taxonomic diversity. *Environmental Science and Pollution Research*. 28(8): 9954–9964.

Guitart, C., Readman, J.W. (2010). Critical evaluation of the determination of pharmaceutical, personal care products, phenolic endocrine disrupters and faecal steroids by GC/MS and PTV-GC/MS in environmental waters. *Analytical Chemistry Acta*. 658(1): 32–40.

Ianora, A., Bentley, M.G., Caldwell, G.S., Casotti, R., Cembella, A.D., Engström-Öst, J., Halsband, C., Sonnenschein, E., Legrand, C., Llewellyn, C.A., Paldavičienë, A., Pilkaityte, R., Pohnert, G., Razinkovas, A., Romano, G., Tillmann, U., Vaiciute, D. (2011). The relevance of marine chemical ecology to plankton and ecosystem function: An emerging field. *Marine Drugs*. 9(9): 1625–1648. http://doi.org/10.3390/md9091625.

Kock, A., Glanville, H.C., Law, A.C., Stanton, T., Carter, L.J., Taylor, J.L. (2023). Emerging challenges of the impacts of pharmaceuticals on aquatic ecosystems: A diatom perspective-review. *Science of the Total Environment*. 878: 162939.

Martin, A. (2020). Maker of a hair-straightening product settles lawsuit. *The New York Times*. Accessed November 10, 2020. Available at: https://www.nytimes.com/2012/03/06/business/brazilian-blowout-agrees-to-a-4-5-million-settlement.html?smid =em-share.

McKnight, U.S., Rasmussen, J.J., Kronvang, B., Binning, P.J., Bjerg, P.L. (2015). Sources, occurrence and predicted aquatic impact of legacy and contemporary pesticides in streams. *Environmental Pollution*. 200: 64–76. https://doi.org/10.1016/j.envpol.2015.02.015.

Miller, T.H., Bury, N.R., Owen, S.F., MacRae, J.I., Barron, L.P. (2018). A review of the pharmaceutical exposome in aquatic fauna. *Environmental Pollution*. 239: 129–146.

Mojiri, A., Bahartoxoeian, M., Zahed, M.A. (2021). The potential of *Chaetoceros muelleri* in bioremediation of antibodies: Performance and optimization. *International Journal of Environmental Research and Public Health*. 18(3): 977.

Molins-Delgado, D.M., Diaz-Cruz, S., Barcelo, D. (2014). Introduction: Personal care products in the aquatic environment. In: Díaz-Cruz, M., Barceló, D. (eds) *Personal Care Products in the Aquatic Environment. The Handbook of Environmental Chemistry, vol 36*. Springer, Cham. https://doi.org/10.1007/698_2014_302

Morganti, P., Paglialunga, S. (2008). EU borderline cosmetic products review of current regulatory status. *Clinics in Dermatology*. 26: 392–397.

Ngqwala, N.P., Muchesa, P. (2020). Occurrence of pharmaceuticals in aquatic environments: A review and potential impacts in South Africa. *South Africa Journal of Science*. 116(7–8): 1–7.

Rizzi, C. (2020). Class action alleges DevaCurl hair products contain several 'known' allergens, irritants. *ClassAction.org*. Accessed October 30, 2020. Available at: https://www.classaction.org/news/class-action-alleges-devacurl-hair-products-contain-several-known-allergens-irritants#embedded-document.

Sotão Neto, B.M.T., Combi, T., Taniguchi, S., Albergaria-Barbosa, A.C.R., Ramos, R.B., Figueira, R.C.L., Montone, R.C. (2020). Persistent organic pollutants (POPs) and personal care products (PCPs) in the surface sediments of a large tropical bay (Todos os Santos Bay, Brazil). *Marine Pollution Bulletin*. 161: 111818. https://doi.org/10.1016/j.marpolbul.2020.111818.

USEPA. (1998). *Guidelines for ecological risk assessment.* Report No. EPA/630/R-95/002F, USEPA, Washington, DC.

Wang, Y., Liu, J., Kang, D., Wu, C., Wu, Y. (2017b). Removal of pharmaceuticals and personal care products from wastewater using algae-based technologies. A review. *Review in Environmental Science and Biotechnology.* 6: 717–735.

Wei, Y., Ding, D., Gu, T., Jiang, T., Qu, K., Sun, J., Cui, Z. (2022). Different responses of phytoplankton and zooplankton communities to current changing coastal environments. *Environmental Research.* 215(Pt 2):114426. http://doi.org/10.1016/j.envres.2022.114426.

Wydro, U., Wołejko, E., Luarasi, L., Puto, K., Tarasevičienė, Z., Jabłońska-Trypuć, A. (2024). A review on pharmaceuticals and personal care products residues in the aquatic environment and possibilities for their remediation. *Sustainability.* 16(1): 169. https://doi.org/10.3390/su16010169.

Xin, X., Huang, G., Zhang, B. (2020). Review of aquatic toxicity of pharmaceuticals and personal care products to algae. *Journal of Hazardous Materials.* 124619.

8 Bacteria and Algae Consortia in Aquatic Ecosystem and the Effects of Personal Care Products on Their Ecological Role

Osikemekha Anthony Anani, Kenneth Kennedy Adama, and Anuwuli Gloria Anani

8.1 INTRODUCTION

The activities of humans from pharmaceutical and cosmetic industries on the ecosystem have generally introduced several priorities of pollutants called emergent into the environment that are accumulative, persistent, recalcitrant, and hazardous (Nowak-Lange et al., 2022). In the last decades, however, there has been a sharp rise in the presence of emergent pollutants whose fates and toxicity of many of them are yet to be ascertained. Most of these materials are novel chemicals that are yet to be evaluated and have been in the ecosystem for some decades now as environmental pollutants (Lapworth et al., 2012).

Most of these pollutants are used in our day-to-day body grooming as PCPs (personal care products) or as additives to improve their quality (Sotão Neto, 2020; Nowak-Lange et al., 2022; Narayanan et al., 2023). Chen et al. (2018) reported that in the international beauty market, humans use 6 of 12 cosmetics per day which cut across toiletries, fragrances, makeup, skincare, and hair care products mainly in France, Japan, and the USA. In these cosmetics used by humans, it is estimated that about twelve (12) thousand chemicals are utilized in their production and about 12% of the chemicals have been known to be safe for the environment and human health (O'Dell et al., 2016; Nowak-Lange et al., 2022). Most of these chemicals are leached into the environment and end up in the aquatic environment via anthropogenic activities. Of course, most of them and their environmental and health impact on the water quality and the aquatic biota like bacteria and algae are unknown (Deblonde et al., 2011; Nowak-Lange et al., 2022). Liu and Wong (2013) reported that most of

DOI: 10.1201/9781003362975-8

the chemicals pass through wastewater and sewage from industrial and domestic activities.

Bacteria and algae play vital roles in the ecosystem structure in any aquatic environment as the basis of the food chain and major transporters of some chemical compounds in the ecosystem (Adetunji and Anani, 2021a, 2021b; Adetunji et al., 2021a, 2021b; Anani et al., 2023; Lobus and Kulikovskiy, 2023). They also control the distribution, bioavailability, and abundance of chemical compounds and certify the effective transfer and accumulation in the sediment and superficial regions in the geological region of the ecosystem. Biogeochemical and food web process (Leonova and Bobrov, 2012; Lobus et al., 2021; Lobus and Kulikovskiy, 2023). However, it has been reported by many studies that emergent pollutants have the potential to upturn the natural status of aquatic systems and thus reduce the water quality and the condition of the species therein (Ávila et al., 2015; Aziz and Ojumu, 2020; Azaroff et al., 2021; Mishra et al., 2023). So there is a possibility of PCPs being leached off into the aquatic environment to cause potential harm to bacteria and algae. Hence, this chapter reviews the potential of personal care products (PCPs) as persistent pollutants in bacteria and algae communities.

8.2 HIGHLIGHTS ON THE ENVIRONMENTAL IMPACTS OF SOME PCPS

Over the last few decades, environmental contaminants associated with some PCPs have posed severe concerns globally, contaminating soil and water systems. The primary source of these contaminants are contaminants of emerging concern (CECs) which include endocrine disruptors (EDCs) and pharmaceuticals and personal care products (PPCPs). The presence of these substances in the ecosystem can trigger a negative impact on human health as well as aquatic bodies, hence, the growing concern and interest in their study and elimination (Anand et al., 2022). Noxious and toxic pharmaceutical wastes and other personal care products belong to the group of xenobiotics of anthropogenic origin, referred to as PPCPs (pharmaceuticals and personal care products). This group could also include food supplements and nutrients (referred to as nutraceuticals), including the various types of cosmetics and their additives, often with antiseptic and antibacterial properties (shampoos, UV blockers, toilet waters, and antiseptics). Other members of this group of compounds also include disinfectants (antibacterial and antiviral) which have become increasingly available since 2020 due to the SARS-CoV-2 pandemic (Phonsiri et al., 2019; Adhikari et al., 2022). The presence of these PCP compounds in the environment is majorly caused by the indiscriminate release of the materials from the environment such as in homes, industries, plant factories, wastewater treatment plants, and healthcare facilities. The United States Environmental Protection Agency listed some groups of PCPs as the most substantial components of these pollutants, and these include metoprolol, atenolol, and carbamazepine (Anand et al., 2022).

Personal care products from different sources show an aggregation of different physicochemical properties depending on the sources of the constituent precursor material. They also possess complex chemical structures and unique stability under different operating conditions which usually makes their environmental impact analysis and estimation as well as the process of their remediation sometimes difficult (Sangion and Gramatica, 2016). However, due to the high number of new personal hygiene products introduced into circulation continually, there have been commensurate difficulties in the environmental assessment of their impacts on the aquatic systems and their remediation on the water body as well as their impact on the environment whose results and output are quite low (Castañeda-Juarez et al., 2022). The pharmaceutical and personal care industries represent one of the fastest-developing human care industries. The factors that have triggered this growth and development in the human care industry include the aging system associated with human society, improvement in the population incomes, improved expenditure on emerging research and development, and increased incidence of civilization and prevalent chronic diseases (González Peña et al., 2021). Moreover, due to the pandemic caused by SARS-CoV-2, there has been significant growth in the use and consumption of some medications and personal care products whose wastes and expired products are often released to the environment indiscriminately. The different PCPs are used to prevent infection and improve human appearance, which in turn has caused an increase in the load of pollutants on bacteria and algae residues in the marine ecosystem (Castañeda-Juárez et al., 2022).

The environmental impact of some PCPs in aquatic environments is significant for the natural environment. These PCPs contain substances or compounds that are persistent and biologically active and, in most cases, can be toxic and capable of disrupting the endocrine system. The other environmental consequences of the presence of these PCPs include damage to the nervous system, the feminization of reproductive functions, or the inhibition of photosynthesis (Ohoro et al., 2019; Motawi et al., 2019). The impact of PCPs on the environment is serious because PCPs in aquatic environments and wastewater are rather detected at low concentrations, and sometimes, their concentrations are somewhat indetectable (Khasawneh and Palaniandy, 2021; Adeleye et al., 2022). While the presence of PCPs in aquatic biota and wastewater in various concentrations is not surprising, a problem arises when they are present in surface and groundwater, which are often sources of drinking water.

8.3 ROLE OF SOME MARINE AND FRESHWATER BACTERIA AND ALGAE IN THE ECOSYSTEM

Lobus and Kulikovskiy (2023) evaluated the geological, chemical, and biological role of phytoplankton-photoautotrophs like the cyanobacteria and the microalgae in an aquatic system. Marine and freshwater phytoplankton have the potential to cycle nutrients, oxygen, and carbon (IV) oxide by using the energy from the sun to fix most of the nutrients needed to be transferred into the ecological food chain or web. More so, they use this energy to maintain their internal environment via homeostasis and use their enzymes which act as cofactors to regulate the activist of emergent pollutants and other related pollutants in the geochemical environment which may have little or more effects on their biosystems. Thus, this may influence their distribution and abundance in their ecozone, then creating an ecological displacement from their natural environment.

Shalaby (2011) investigated the role of algae in contributing to the environmental health condition in any aquatic ecosystem. Algae produce significant compounds that are well known as secondary bio metabolites which are made specifically towards the end growth phase of their development elicited by certain environmental stress like drought, high salinity, and increase or change in their internal environment. Some of these metabolites are unsaturated fatty acids, polysaccharides, pigments of phycobiliprotein, compounds of phenolics, and carotenoids which have shown strong antimicrobial, anticancer, and antioxidant activities against certain chemicals and biological organisms in the environment. Interestingly, algae have been utilized in the pharmaceutical, cosmetic, and food industries which has also served as a major point to generate funds in the current global market space. In addition, the role of macroalgae and bacteria have significantly influenced the aquaculture sector because they are used to provide major basic nutrients for so many organisms in the aquatic ecosystem like fatty acids, sterols, nitrogen-related compounds, and vitamins. Shalaby (2011) reported that over 45.71 mmt (million metric tons) cost $56.47 billion of algae products in the year 2000 was used for domestic aquaculture production. This means that algae/microalga remains one of the widely utilized biomass for aquaculture production and renewable energy (gasification and combustion).

Maréchal (2019) assessed the expectations and challenges of freshwater and marine plants. The authors reported that in recent years, photosynthetic organisms in the aquatic environment have contributed to several ecosystem services that have benefited the basic food chain and food webs respectively. Freshwater and marine plants like algae have generally been affected by climate change and pollutants that are emergent or otherwise. They have been stressed beyond their natural state. Understanding their physiology and genomics will strike a balance in studying their potential adaption to their immediate under-stressed ecozone. This will lead to a well-balanced and developed algae-built economy.

Hamidi et al. (2019) examined the role of microalgae and bacteria as potential biotechnological tools in the environment. Microalgae and bacteria have been found to possess natural biological compounds with strong antioxidant potentials in the deterrence of ROS (reactive oxygen species) formation. Because of the very harsh chemical, physical, and biological settings the marine prokaryotic and eukaryotic species have to deal with, they have developed certain techniques like the formation of antioxidants like catalase, dismutase, superoxide, peptides, exopolysaccharides, and carotenoids to manage or adapt to their immediate environment. Hence, bacteria and algae have been utilized for the production of cosmeceuticals, nutraceuticals, and pharmaceutical products because of the basic chattels they possess.

Dell' Anno et al. (2021) investigated the role of microalgae, fungi, and bacteria sourced from the benthic sediment region of marine ecosystems as potential bioremediation tools to remediate PHCs (petroleum hydrocarbons). PHCs are widespread organic pollutants in marine zones that have been reported to have varied health and environmental effects. However, there is a need to call for a sustainable solution via some mitigation approach to protect the environment and its natural entities, biological, chemical, and physical alike. This can be a reclamation of the marine ecosystem by using a microbial approach—microalgae, fungi, and bacteria to reclaim the polluted environment. This method has been adjudged to be green, clean, eco-friendly, cheap, and eco-compatible, and the organisms have been proven to have the potential to degrade pollutants that are recalcitrant like the emergent ones. However, their level of degradation varies on the bioavailability and chemical structure of the pollutants and the settings in the environment.

Ramanan et al. (2016) evaluate the bacteria-algae relations and their application in the control of environmental pollutants. In the early evolution period till date, bacteria-algae coexisted. Their influence in the ecosystem, like the marine environment, occurs because lichens form several parasitic and mutualistic bio-associations. Ramanan et al. (2016) reported that studies have indicated that bacteria and algae affect themselves synergistically, affecting each other metabolically and physiologically. An example is in an interaction called rosewater algae. This association is found almost everywhere in any aquatic environment thus influencing the primary product in any food chain structure. A few decades ago, algae received special interest in the scientific world for manufacturing and engineering exploitation because of their potential as food, biofuels, and for medical uses. Ramanan et al. (2016) also noted that bacteria on the other hand can stimulate the growth of algae. However, there is a need to know the biological and physiological interaction between these two organisms to utilize and integrate them for industrial purposes.

8.4 THE EFFECTS OF REENGINEERING BACTERIA AND ALGAE TO EMERGENT POLLUTANTS

Reengineering microorganisms, such as bacteria, fungus, algae, and yeasts to remove different pollutants from aquatic systems, has been recognized as an alternative eco-friendly green technology. In recent decades, the broad range of bacteria and algae communities in emergent pollution systems has continued to attract considerable interest. The process is a means of reducing the accumulation of organic waste pollutants and mitigating the scourge of environmental contamination. Numerous novel and reengineered algae and bacteria communities have been investigated to determine their efficacy as a pollutant containment material in aquatic ecosystems. These approaches would contribute to the overall sustainability by minimizing the costs and carbon emissions associated with algae to contribute to the much-touted, circular, zero-carbon global economy.

Emergent pollutants (EP) can be considered as synthetic organic chemicals that include pharmaceuticals, personal care products (PCPs), herbicides, pesticides, and flame retardants, whose presence in the environment is of concern due to their potential risks to ecosystems and human health at environmentally relevant concentrations (Petrie et al., 2015; Tran et al., 2018; Sutherland and Ralph, 2019). There is increasing concern over the presence of EPs in aquatic systems. With climate change and expanding populations, accumulating EPs due to increased water reuse could lead to unpredictable long-term consequences for humans and the environment (Martinez-Piernas et al., 2018).

The process of reengineering bacteria and algae to emergent pollutants is still in its infancy. However, algae and bacteria biodegradation provide one of the most promising technologies to transform, neutralize, or eliminate emergent pollutants from aquatic systems. Unlike other remediation techniques, such as activated carbon adsorption filters, which simply concentrate the emergent pollutants and remove them from one environment to another environment, biodegradation involves the transformation of complex compounds into simpler breakdown molecules through catalytic metabolic degradation (Sutherland and Ralph, 2019). Bacteria and algae degradation of emergent pollutants can occur via two main mechanisms. The first mechanism involves direct metabolic degradation of the emergent pollutant by the bacteria and algae (Tiwari et al., 2017). The second mechanism involves indirect, or co-metabolism, where the emergent pollutant is degraded by enzymes that are catalyzing other substrates present (Tiwari et al., 2017). Both bacteria and algae possess a large number of enzymes that play critical roles in cellular protection through the deactivation and/or degradation of a range of organic compounds that induce cellular stress (Wang et al., 2019).

8.5 LEGAL VIEWS OR REGULATION ISSUES ON BIOENGINEERING (BACTERIA AND ALGAE)—MICROALGAE

Kumar et al. (2020) reported several legal issues on the bioengineering of bacteria and algae in industrial applications. Due to the exploitation of bacteria and algae, specifically microalgae, in the pharmaceuticals and nutraceutical industries because of their bioactive substances, the interest to boost many strains of organisms using engineering means has been heralded recently. However, their application has sparked various concerns because of some limitations in the bioengineering processes, the unforeseen health impacts they portend for the end users, and the possible environmental and genetic drift it causes in the immediate ecosystem. This may also cause some biosafety and future regulation issues if not looked into properly.

Kumar et al. (2020) reported that in some parts of the globe, strict policies/laws/regulations are needed to check the recombinant/transgenic microalgae to ensure total compliance during the modification of the genetic traits to improve their vigor for commercial utilization. Kumar et al. (2020) also reported that Algenol, an engineering company in the US (Florida), was permitted to use genetically modified cyanobacteria for in- and outdoor agronomy. However, the CBD (Convention on Biological Diversity) in 2015 findings has shown that these bioengineering microorganisms have been found to carry noxious materials that may portend health and environmental concerns (https://www.cbd.int/ts/cbd-ts-82-en.pdf). Under the TSCA (Toxic Substances Control Act) of the US EPA which is an outed tool in the assessment or regulation of toxic substances organisms, risk may arise from this bio-invention. The same act requires companies to file notice on MCA (Microbial Commercial Activity) before any commercialization of novel genetically modified species. Nethravathy et al. (2019) reported that to date, reports on the cultivation of outdoor genetically modified microorganisms have not been reported because of the unexpected and practicable hazards linked to open cultivation of the organisms in the wild. Cultivation may portend several hazards which may become uncontrollable, thus polluting natural species, competing with critical resources in the ecosystem, and resulting in a mis-normal breeding of sexually well-matched species. This becomes one of the critical ethical and legal issues to combat in the future when there are any possible negative episodes.

Hence, strict biosecurity measures are needed to safeguard the transportation and importation of strains of microorganisms and algae that may have serious ecosystem, environmental, and genetic modification of natural species in the wild (Campbell, 2011; Kumar et al., 2020). Thus, to manage this scenario, strict regulation via monitoring of the cultivation and the handling processes with environmental and health tools should be the integral biosafety standards. Stakeholders and the government should make sure that these biosafety tools are adhered to with fines or punishment for defaulters.

8.6 DEVELOPMENT AND PROSPECTS IN A GREEN ECOSYSTEM

Wang et al. (2023) in a study demonstrated the need to use quality green technology that is eco-friendly as a policy tool in river biomonitoring. The results of the study clearly showed that a more balanced direction is needed for a scientific and theoretical establishment for a comprehensive assessment of the model before its general application, especially in the Delta River of the Yangtze region. This is because the model only provides a specific assessment of the achievements and process development without the social, economic, and ecological calculation of the ecosystem in view. However, the major application of the model is as an indicator system and an EGI (Ecological Green Integration) that covers production, life, and the ecology within the Delta River of the Yangtze region. Based on this, the model can only be recommended to capture restricted data.

Su (2022) evaluated the impact of green technology on the ecosystem and a prospective economy. In this study, about 30 provinces of China were investigated using different methods: PVAR (panel vector autoregressive model), CCM (coupling coordination model), and the EWM (entropy weight method) to ascertain the green economy, innovation in technology, and the extent of the development in the ecosystem from 2005 to 2016. The results from this study showed a comprehensive advancement in the levels of the green economy, innovation in technology, and the extent of the development in the ecosystem in the provinces investigated respectively. Based on the findings, it was noticed that by improving the levels of the ecosystem and technology of the various provinces investigated, the green economy of the various provinces will bloom.

Hassan et al. (2023) looked at the ecosystem footprint in the determination of green growth using natural resources, innovation techniques, and ICT (information technology). The relationship between ecosystem footprint and green growth is linked with important ecological implications. This study deployed the CS-ARDL (cross-sectional-autoregressive-distributed lag) technique to determine the short and long-run estimates of a perfect ecosystem. The results from this study revealed that there were reductions in the green growth of natural resources, ecosystems, and ICT. However, more findings showed that the natural resources had more efficacy than the ecosystem footprint. But then there was a reduction in the ecosystem footprint as a result of the efficiency of the natural resources and the ICT. The authors recommend that ecofriendly technology is needed to sustain the green process of development which is also necessary to aid combat and protect the environment against unforeseen pollution.

Houssam et al. (2023) assessed the importance of a sustainable and green economy in developing nations. The green economy has been heralded to be one of the major tools in both developed and developing countries of the world. Thus, this study applied the GLS (generalized least square) method to assess the accomplishment of the GGE (global green economy). Findings from the study revealed that a positive correlation exists between the green economy and the gross domestic products. Meanwhile, a negative correlation existed between the poverty level and the green economy as in the developing countries of the world. By implication, this study struck a balance between the developed and developing nations in the sense that for an ecosystem to thrive fully, a balanced economy of a nation is needed to aid the required changes needed to structure well as against the case when the economy is stricken with struggling of gross domestic products. Hence, countries with low gross domestic products may not have the wherewithal to cater for episodes of climate change, ecosystem degradation, and loss of biodiversity, where there is incidence of constant and uncontrolled discharge of pollution into the environment like the emergent ones. In light of this, there needs to be a paradigm shift to balance research and development in developing countries to improve the aspect of green technology and a stable economy. Also, the use of renewable energy is encouraged to cut the incidence of global warming and climate change via the production of eco-friendly biofuels and biodiesels to replace fossil fuels.

8.7 CONCLUSION

The activities of humans from pharmaceutical and cosmetic industries on the ecosystem have generally introduced several priorities of pollutants called emergent into the environment that are accumulative, persistent, recalcitrant, and hazardous. This chapter reviews the potential of personal care products (PCPs) as persistent pollutants in bacteria and algae communities.

These groups could include food supplements and nutrients (referred to as nutraceuticals), including the various types of cosmetics and their additives, often with antiseptic and antibacterial properties (shampoos, UV blockers, toilet waters, and antiseptics). Highlights on the environmental impacts of some PCPs showed that the presence of these substances in the ecosystem can trigger negative impacts on human health as well as aquatic bodies, hence, the growing concern and interest in their study and elimination. These PCPs contain substances or compounds that are persistent and biologically active and, in most cases, can be toxic and capable of disrupting the endocrine system. The other environmental consequences of the presence of these PCPs include damage to the nervous system, the feminization of reproductive functions, or the inhibition of photosynthesis.

Marine and freshwater phytoplankton have the potential to cycle nutrients, oxygen, and carbon (IV) oxide by using the energy from the sun to fix most of the nutrients needed to be transferred into the ecological food chain or web. The role of some marine and freshwater bacteria and algae in the ecosystem is very significant in structuring the ecosystem. Algae produce significant compounds that are well-known as secondary bio metabolites which are made specifically towards the end growth phase of their development elicited by certain environmental stress like drought, high salinity, and increase or change in their internal environment. Some of these metabolites are unsaturated fatty acids, polysaccharides, pigments of phycobiliprotein, compounds of phenolics, and carotenoids which have shown strong antimicrobial, anticancer, and antioxidant activities against certain chemical and biological organisms in the environment. Freshwater and marine plants like algae have generally been affected by climate change and pollutants that are emergent or otherwise.

The effects of reengineering bacteria and algae to emergent pollutants have sparked great interest in the field of biotechnology recently. Reengineering microorganisms such as bacteria, fungus, algae, and yeasts to remove different pollutants from aquatic systems has been recognized as an alternative eco-friendly green technology. The process of reengineering bacteria and algae to emergent pollutants is still in its infancy. However, algae and bacteria biodegradation provide one of the most promising technologies to transform, neutralize, or eliminate emergent pollutants from aquatic systems.

Regulation problems on bioengineering (bacteria and algae), specifically microalgae, have been an issue of discussion in the bioengineering fields of recent. Due to the exploitation of bacteria and algae, specifically microalgae, in the pharmaceuticals and nutraceutical industries because of their bioactive substances, the interest to boost many strains of organisms using engineering means has been heralded recently. However, their application has sparked various concerns because of some limitations in the bioengineering processes, the unforeseen health impacts they portend for the end users, and the possible environmental and genetic drift it causes in the immediate ecosystem. Hence, strict biosecurity measures are needed to safeguard the transportation and importation of strains of microorganisms and algae that may have serious ecosystem, environmental, and genetic modification of natural species in the wild. Stakeholders and the government should make sure that these biosafety tools are adhered to with fines or punishment for defaulters.

The impact of green technology on the ecosystem and a prospective economy is one of the ways to balance any perturbed ecosystem. The development and prospects in a green ecosystem are very crucial because natural resources are harnessed to improve the efficacy of the ecosystem footprint which is also necessary to aid combat and protect the environment against unforeseen pollution, a balanced economy, and control climate change and biodiversity loss.

REFERENCES

Adeleye, A.S., Xue, J., Zhao, Y., Taylor, A.A., Zenobio, J.E., Sun, Y., Han, Z., Salawu, O.A., Zhu, Y. (2022). Abundance, fate, and effects of pharmaceuticals and personal care products in aquatic environments. *Journal of Hazardous Materials*. 424: 127284.

Adetunji, C.O., Anani, O.A. (2021a). Recent advances in the application of genetically engineered microorganisms for microbial rejuvenation of contaminated environment. In *Microbial Rejuvenation of Polluted Environment*. Springer. http://doi.org/10.1007/978-981-15-7459-7_14

Adetunji, C.O., Anani, O.A. (2021b). Plastic-eating microorganisms: Recent biotechnological techniques for recycling of plastic. In *Microbial Rejuvenation of Polluted Environment*. Springer. http://doi.org/10.1007/978-981-15-7447-4_14

Adetunji, C.O., Anani, O.A., Egbuna, C. (2021a). *Utilization of Biosurfactant Derived from Beneficial Microorganisms as Sustainable Bioremediation Technology for the Management of Contaminated Environment: Panacea for a Healthy Planet.* Apple Academic Press, Inc. 15.

Adetunji, C.O., Anani, O.A., Panpatte, D. (2021b). Mechanism of actions involved in sustainable ecorestoration of petroleum hydrocarbons polluted soil by the beneficial microorganism. In Panpatte, D.G., Jhala, Y.K. (eds) *Microbial Rejuvenation of Polluted Environment. Microorganisms for Sustainability*, vol 26. Springer, Singapore. https://doi.org/10.1007/978-981-15-7455-9_8

Adhikari, S., Kumar, R., Driver, E.M., Perleberg, T.D., Yanez, A., Johnston, B., Halden, R.U. (2022). Mass trends of parabens, triclocarban and triclosan in Arizona wastewater collected after the 2017 FDA ban on antimicrobials and during the COVID-19 pandemic. *Water Research.* 222: 118894.

Anand, U., Adelodun, B., Cabreros, C., Kumar, P., Suresh, S., Dey, A., Ballesteros, F., Bontempi, E. (2022). Occurrence, transformation, bioaccumulation, risk and analysis of pharmaceutical and personal care products from wastewater: A review. *Environmental Chemistry Letters.* 20: 3883–3904.

Anani, O.A., Inobeme, A., Osarenotor, O., Olisaka, N.F., Aidonojie, A.P., Olatunji, O.E., Habib, A.I. (2023). Application of microorganisms as Biofactories to produce biogenic nanoparticles for environmental cleanup: Currents advances and challenges. *Current Nanoscience.* 19. https://doi.org/10.2174/1573413719666221219164613.

Ávila, C., Bayona, J.M., Martín, I., Salas, J.J., García, J. (2015). Emerging organic contaminant removal in a full-scale hybrid constructed wetland system for wastewater treatment and reuse. *Ecology and Engineering.* 80: 108–116.

Azaroff, A., Monperrus, M., Miossec, C., Gassie, C., Guyoneaud, R. (2021). Microbial degradation of hydrophobic emerging contaminants from marine sediment slurries (*Capbreton Canyon*) to pure bacterial strain. *Journal of Hazardous Materials.* 402: 123477.

Aziz, M., Ojumu, T. (2020). Exclusion of estrogenic and androgenic steroid hormones from municipal membrane bioreactor wastewater using UF/NF/RO membranes for water reuse application. *Membranes.* 10: 37.

Campbell, M.L. (2011). Assessing biosecurity risk associated with the importation of non-indigenous microalgae. *Environmental Research.* 111: 989–998. https://doi.org/10.1016/j.envres.2011.02.004.

Castañeda-Juárez, M., Linares-Hernández, I., Martínez-Miranda, V., Teutli-Sequeira, E.A., Castillo-Suárez, L.A., Sierra-Sánchez, A.G. (2022). SARS-CoV-2 pharmaceutical drugs: A critical review on the environmental impacts, chemical characteristics, and behavior of advanced oxidation processes in water. *Environmental Science and Pollution Research.* 29: 67604–67640.

Chen, X., Sullivan, D.A., Sullivan, A.G., Kam, W.R., Liu, Y. (2018). Toxicity of cosmetic preservatives on human ocular surface and adnexal cells. *Experiment and Eye Research.* 170: 188–197.

Deblonde, T., Cossu-Leguille, C., Hartemann, P. (2011). Emerging pollutants in wastewater: A review of the literature. *International Journal of Hygiene and Environmental Health.* 214: 442–448.

Dell' Anno, F., Rastelli, E., Sansone, C., Brunet, C., Ianora, A., Dell' Anno, A. (2021). Bacteria, fungi and microalgae for the bioremediation of marine sediments contaminated by petroleum hydrocarbons in the omics era. *Microorganisms.* 9(8): 1695. https://doi.org/10.3390/microorganisms9081695.

González Peña, O.I., López Zavala, M.Á., Cabral Ruelas, H. (2021). Pharmaceuticals market, consumption trends and disease incidence are not driving the pharmaceutical research on water and wastewater. *International Journal of Environmental Research and Public Health*. 18: 2532.

Hamidi, M., Kozani, P.S., Kozani, P.S., Pierre, G., Michaud, P., Delattre, C. (2019). Marine bacteria versus microalgae: Who is the best for biotechnological production of bioactive compounds with antioxidant properties and other biological applications? *Marine Drugs*. 18(1): 28. https://doi.org/10.3390/md18010028.

Hassan, A., Yang, J., Usman, A., Bilal, A., Ullah, S. (2023). Green growth as a determinant of ecological footprint: Do ICT diffusion, environmental innovation, and natural resources matter? *PLoS ONE*. 18(9): e0287715. https://doi.org/10.1371/journal.pone.0287715.

Houssam, N., Ibrahiem, D.M., Sucharita, S., El-Aasar, K.M., Esily, R.R., Sethi, N. (2023). Assessing the role of green economy on sustainable development in developing countries. *Heliyon*. 9(6): e17306. https://doi.org/10.1016/j.heliyon.2023.e17306.

Khasawneh, O.F.S., Palaniandy, P. (2021). Occurrence and removal of pharmaceuticals in wastewater treatment plants. *Process, Safety, and Environmental Protection*. 150: 532–556.

Kumar, G., Shekh, A., Jakhu, S., Sharma, Y., Kapoor, R., Sharma, T.R. (2020). Bioengineering of microalgae: Recent advances, perspectives, and regulatory challenges for industrial application. *Frontier in Bioengineering and Biotechnology*. 8: 914. https://doi.org/10.3389/fbioe.2020.00914.

Lapworth, D.J., Baran, N., Stuart, M.E., Ward, R.S. (2012). Emerging organic contaminants in groundwater: A review of sources, fate and occurrence. *Environmental Pollution*. 163: 287–303.

Leonova, G.A., Bobrov, V.A. (2012). *Geochemical Role of Plankton of Continental Water Bodies in Siberian in Concentration and Biosedimentation of Microelements*. Geo, Novosibirsk, Russia.

Liu, J.L., Wong, M.H. (2013). Pharmaceuticals and personal care products (PPCPs): A review on environmental contamination in China. *Environmental International*. 59: 208–224.

Lobus, N.V., Kulikovskiy, M.S. (2023). The co-evolution aspects of the biogeochemical role of phytoplankton in aquatic ecosystems: A review. *Biology (Basel)*. 1: 92. https://doi.org/10.3390/biology12010092.

Lobus, N.V., Kulikovskiy, M.S., Maltsev, Y.I. (2021). Multi-element composition of diatom *Chaetoceros spp*. from natural phytoplankton assemblages of the Russian Arctic seas. *Biology*. 10: 1009. https://doi.org/10.3390/biology10101009.

Maréchal, E. (2019). Marine and freshwater plants: Challenges and expectations. *Frontier in Plant Science*. 10: 1545. https://doi.org/10.3389/fpls.2019.01545.

Martinez-Piernas, A.B., Plaza-Bolanos, P., Garcia-Gomez, E., Fernandez-Ibanez, P., Aguera, A. (2018). Determination of organic micro-contaminants in agricultural soils irrigated with reclaimed wastewater: Target and suspect approaches. *Analytic in Chemistry Acta*. 1030: 115–124.

Mishra, R.K., Mentha, S.S., Misra, Y., Dwivedi, N. (2023). Emerging pollutants of severe environmental concern in water and wastewater: A comprehensive review on current developments and future research. *Water-Energy Nexus*: 74–95.

Motawi, T.K., Ahmed, S.A., El-Boghdady, N.A., Metwally, N.S., Nasr, N.N. (2019). Protective effects of betanin against paracetamol and diclofenac induced neurotoxicity and endocrine disruption in rats. *Biomarker as Biochemical Indicators Exposure Response Susceptibility Chemistry*. 24: 645–651.

Narayanan, M., Kandasamy, S., Lee, J., Barathi, S. (2023). Microbial degradation and transformation of PPCPs in aquatic environment: A review. *Heliyon*. 9: e18426.

Nethravathy, M., Mehar, J.G., Mudliar, S.N., Shekh, A.Y. (2019). Recent advances in microalgal bioactives for food, feed, and healthcare products: Commercial potential, market space, and sustainability. *Comprehensive Review in Food Science and Food Safety*. 18: 1882–1897. https://doi.org/10.1111/1541-4337.12500.

Nowak-Lange, M., Niedziałkowska, K., Lisowska, K. (2022). Cosmetic preservatives: Hazardous micropollutants in need of greater attention? *International Journal of Molecular Science*. 23: 14495. https://doi.org/10.3390/ijms232214495.

O'Dell, L.M., Sullivan, A.G., Periman, L.M. (2016). Beauty does not have to hurt. *Advanced Ocular Care*: 42–47.

Ohoro, C.R., Adeniji, A.O., Okoh, A.I., Okoh, O.O. (2019). Distribution and chemical analysis of Pharmaceuticals and Personal Care Products (PPCPs) in the environmental systems: A review. *International Journal of Environmental Research and Public Health*. 16: 3026.

Petrie, B., Barden, R., Kasprzyk-Hordern, B. (2015). A review on emerging contaminants in wastewaters and the environment: Current knowledge, understudied areas and recommendations for future monitoring. *Water Research*. 72: 3–27.

Phonsiri, V., Choi, S., Nguyen, C., Tsai, Y.-L., Coss, R., Kurwadkar, S. (2019). Monitoring occurrence and removal of selected pharmaceuticals in two different wastewater treatment plants. *SN Applied Science*. 1: 798.

Ramanan, R., Kim, B.-H., Cho, D.-H., Oh, H.-M., Kim, H.-S. (2016). Algae–bacteria interactions: Evolution, ecology and emerging applications. *Biotechnology Advances*. 34(1): 14–29.

Sangion, A., Gramatica, P. (2016). Ecotoxicity interspecies QAAR models from daphnia toxicity of pharmaceuticals and personal care products. *SAR QSAR Environmental Research*. 27: 781–798.

Shalaby, E.A. (2011). Algae as promising organisms for environment and health. *Plant Signal Behavior*. 9: 1338–1350. https://doi.org/10.4161/psb.6.9.16779.

Sotão Neto, B.T.M., Combi, T., Taniguchi, D., Albergaria-Barbosa, Ana C.R., Ramos, R.B., Figueira, R.C.L., Montone, R.C. (2020). Persistent organic pollutants (POPs) and personal care products (PCPs) in the surface sediments of a large tropical bay (Todos os Santos Bay, Brazil). *Marine Pollution Bulletin*. 161: 111818.

Su, L. (2022). The impact of coordinated development of ecological environment and technological innovation on green economy: Evidence from China. *International Journal of Environmental Research and Public Health*. 19(12): 6994. https://doi.org/10.3390/ijerph19126994.

Sutherland, D.L., Ralph, P.J. (2019). Microalgal bioremediation of emerging contaminants—Opportunities and challenges. *Water Research*. 164: 114921.

Tiwari, B., Sellamuthu, B., Ouarda, Y., Drogui, P., Tyagi, R.D., Buelna, G. (2017). Review on fate and mechanism of removal of pharmaceutical pollutants from wastewater using biological approach. *Bioresources Technology*. 224: 1–12.

Tran, N.H., Reinhard, M., Gin, K.Y. (2018). Occurrence and fate of emerging contaminants in municipal wastewater treatment plants from different geographical regions-a review. *Water Research*. 133: 182–207.

Wang, C., Dong, D., Zhang, L., Song, Z., Hua, X., Guo, Z. (2019). Response of freshwater biofilms to antibiotic florfenicol and ofloxacin stress: Role of extracellular polymeric substances. *International Journal of Environmental Research and Public Health*. 16: 715.

Wang, W., Chen, L., Yan, X. (2023). Evaluation of ecological green high-quality development based on network hierarchy model for the demonstration area in Yangtze River Delta in China. *Front Public Health*. 11: 1159312. https://doi.org/10.3389/fpubh.2023.1159312.

9 Effects of Personal Care Products on Surface Water Floating Organisms

Ebere Mary Eze, Oke Aruoren,
Joshua Othuke Orogu, Odiri Ukolobi,
Peter Mudiaga Etaware, and Joel Okpoghono

9.1 INTRODUCTION

The types of pollutants found on water surfaces vary depending on where they come from, but most of the common ones are nutrients, pathogenic microbes, solid wastes, inorganic and organic compounds, and micropollutants. The discharge of these contaminants into the environment has a detrimental effect on the ecology, human health, and the economy (Iyer et al., 2021).

Most people use personal care products daily, so there is more of a demand scope for wastes both solid wastes and wastewater generated from their use. Personal care product wastewater is carried from household sinks and drains to wastewater treatment. On the other hand, homes that are not linked to the sewerage system directly discharge their raw wastewater into the environment via open canals that eventually flow into the closest surface waters (Orii & Kannan, 2008).

In addition, they eventually contaminate groundwater as well as any soil surfaces they come into contact with along the journey. It is significant to remember that wastewaters globally are released into the environment untreated. The evidence presented leads one to the conclusion that a significant portion of the compounds contained in personal care products that are present in wastewater instantly contaminate the natural environment's soil and water surfaces. Personal care products (PCPs) compounds can have both direct and indirect effects, including disruption of the endocrine system, impact on biochemical processes, development of antimicrobial resistance, and bioaccumulation of the compounds in non-target organisms (Frédéric & Yves, 2014). According to Collado et al. (2014), the aquatic environment has become overpopulated with both active and inactive metabolites as a result of the inappropriate disposal of PCP molecules. When non-target organisms bioaccumulate these PCP chemicals in surface waters, they may make their way up the food chain.

Environmental pollution is caused by a multitude of sources, including runoff streams from farms, discharge effluents from hospitals and industrial facilities, leaching from residential septic tanks, and improper disposal of PCPs (Fenech et al.,

DOI: 10.1201/9781003362975-9

2013; Iglesias et al., 2014). Practically speaking, PCPs can go into the environment via one of two methods:

(1) Manufacturing flaws or disposal and the placing of unneeded (or outdated) medication in trash cans, sinks, or toilets; these items may be burned or disposed of in landfills.
(2) Effluents and excretion that result from ineffective plants that treat wastewater.

9.2 CLASSES OF PCPs

PCPs fall into several categories, such as UV filters (methylbenzylidene camphor), perfumes (musks), insect repellents (DEET), disinfectants (triclosan), and preservatives (parabens).

9.2.1 Disinfectants

TCS and TCC are two of the top ten organic wastewater chemicals that are frequently discovered, both in terms of frequency and concentration (Kolpin et al., 2002). TCS has detrimental ecological impacts when it is present in aquatic environments: it exhibits toxicity to algae species and alters the benthic bacterial community's composition (fostering cyanobacteria over algae).

9.2.2 Fragrances (Musk)

Fragrances are possibly the PCP class that has been studied the most, and they are thought to have environmental pollutants that are widely present (Daughton & Ternes, 1999). Synthetic musk perfumes are the most widely used type. Fragrances known as synthetic musks are found in a variety of goods, such as detergents, soaps, and deodorants. One type of synthetic musk is nitro musk, which was first. The two most commonly used nitro musks are musk xylene (MX) and musk ketone (MK); musk ambrette (MA), muskmoskene (MM), and musk tibetene (MT) are used less frequently (Daughton & Ternes, 1999). However, because of their environmental permanence and probable harm to aquatic organisms, nitro musks are gradually being phased out (Daughton & Ternes, 1999). These days, nitro musks are less common while polycyclic musks are more common; celestolide (ABDI), galaxolide (HHCB), and toxalide (AHTN) are the polycyclic musks that are most commonly used, whereas traseloide (ATII), phantolide (AHMI), and cashmeran (DPMI) are the ones that are used least frequently (Daughton & Ternes, 1999).

While comparatively non-toxic to fish, HHCB and AHTN are harmful to aquatic invertebrates at ppb to low ppm levels. Additionally, during extended exposure times, invertebrates are more susceptible to polycyclic musks than fish. Up to eight more smells have been identified in surface water, including skatole, indole, isoborneol, ethyl citrate, camphor, D-limonene, and acetophenone.

However, all scents—aside from ethyl citrate—have been discovered in a limited amount of samples (Kolpin et al., 2002). In water surfaces across the United

States, ethyl citrate, a tobacco ingredient, has been regularly found (Kolpin et al., 2002).

9.2.3 Insect Repellants

N, N-diethyl-m-toluamide (DEET) is commonly seen in water surfaces, and it is the most useful and active ingredient found in insect repellants (Costanzo et al., 2007). Despite being widely present in water surfaces and very resistant to degradation, DEET is relatively persistent in aquatic environments. However, there are no documented studies that have investigated the long-term toxicity of DEET exposure to aquatic life.

9.2.4 Preservatives

Alkyl-p-hydroxybenzoates, or parabens, are antimicrobial preservatives found in food, medicine, cosmetics, and hygiene (Daughton & Ternes, 1999). Nowadays, parabens come in seven varieties (propyl, isopropyl, benzyl, ethyl, isobutyl, butyl, and methyl). Over 7,000 kg of parabens were used in toiletries and cosmetics alone in 1987 (Soni et al., 2001), and for the past 20 years, it has been anticipated that this amount will rise. More than increasing chain length, chlorination also substantially increases the toxicity of parabens to bacteria. It seems that aquatic organisms may be negatively impacted by benzyl, butyl, and propylparaben according to the scant environmental concentration and toxicity data. According to Dobbins et al. (2009), parabens only provide a little risk to aquatic life, but they can cause mild estrogenic reactions in certain cases. Specifically, benzyl, butyl, and propylparaben can cause these reactions. These findings proposed that aquatic organisms continuously exposed to parabens may have five possible impacts. These findings also suggest that aquatic organisms exposed to parabens regularly may have negative impacts. However, since effect concentrations are usually 1,000 times higher than surface water observations, preliminary data on environmental concentrations suggest only a modest risk to aquatic life.

9.2.5 UV Filters

Growing concern over the outcome of ultraviolet (UV) radiation in humans has caused an increased usage of UV filters. UV filters are used in cosmetics and sunscreen products to protect from UV radiation and can either be organic (absorb UV radiation, e.g. methyl benzylidene camphor) or inorganic micro pigments (reflect UV radiation, e.g. ZnO, TiO_2). Sunscreens and cosmetics typically contain three to eight different UV filters, which can account for more than 10% of the bulk of the product (Dobbins et al., 2009).

9.2.6 Sources of PCPs Pollution in Surface Water

PCPs are widely used and one of the ways that human activity contributes to environmental contamination. Irrespective of the fact that alkylphenolpolyethoxylates

(APEOs) decompose into alkylphenol, PCPs commonly use alkylphenolpolyethoxylates as surfactants, such as nonylphenol and octylphenolethoxylates. APEOs are found in several PCPs, disinfectants, detergents, and surface cleaners (Dodson et al., 2012). Additionally, certain PCPs contain antimicrobials due to their aseptic nature. The most often used PCP antibacterial agents are orthophenylphenol, triclosan, triclocarban, and 1,4-dichlorobenzene. Items like these types of antimicrobials are commonly found in detergents, soaps, deodorants, and toothpaste, (Dodson et al., 2012; Allmyr et al., 2006).

Bisphenols are utilized in the production of plastics, such as epoxy resins and polycarbonates. These are not planned components of PCPs; their presence is caused by their migration from plastic containers or degradation (Lu et al., 2018). The most prevalent PCP, bisphenol A, was found in shampoos, lotions, soaps, and detergents, as well as sunscreens, conditioners, shave creams, and nail polishes, as stated by Dodson et al. (2012).

Because of their conditioning and spreading qualities, cyclic volatile methyl siloxanes—also known as cyclosiloxanes—and linear siloxanes are used in PCPs, including sunscreen, infant products, body washes, lotion, hair care products, shaving creams, and cleansers, as well as makeup (Dodson et al., 2012). More than 252 various PCPs in Canada were discovered to have low molecular weight cyclosiloxanes (Wang et al., 2009). Ammonia molecules known as ethanolamines are utilized in cosmetics and PCPs due to their role as surfactants and their antistatic, emulsifying, conditioning, foaming, and viscosity-increasing properties (Fiume et al., 2017).

Perfumes and fragrances are made up of different substances. PCPs are scented with between 50 and 300 different compounds. These chemicals include, but are not limited to, heterocyclics, acetals, pyrazines, amides, carboxylic acids, nitriles, aldehydes, alcohols, coumarins, dioxanes, amines, epoxides, esters, musks, ethers, ketones, lactones, phenols, hydrocarbons, pyrans, quinolines, or Schiff's bases according to Bickers et al. (2003).

Numerous PCPs, including fabric softeners, detergents, soaps, and cleansers, contain a series of fragrances (Bickers et al., 2003). Perfumes, lotions, body creams, deodorants, face cleansers, and sunscreen are examples of PCPs that typically have higher concentrations of fragrances, even artificial ones (Dodson et al., 2012; Reiner & Kannan, 2006).

Glycol ethers are good cleaning agents used in face lotions, cleansers, sunscreen, and shaving creams due to their lipophilic and hydrophilic qualities (Dodson et al., 2012).

United States introduced N, N-diethyl-m-toluamide, or DEET in 1946 as an insect repellent to shield its soldiers from mosquito bites. The Environmental Protection Agency and the Centers for Disease Control and Prevention recommend DEET as the primary insect repellent—mosquito repellent (Patel et al., 2016). Other substances, commonly employed as insect repellents, besides bay-repel, are indole and piperonylbutoxide.

PCPs contain parabens as a preservative. PCPs have long included phthalates, also known as phthalic acid esters, as plasticizers and additives. Diethyl, dimethyl, di-isobutyl, di-n-butyl, and di(2-ethylhexyl) phthalates are the most commonly used. These are frequently found in skin cleansers, nail polishes, hair products, infant items, cosmetics, and scents.

Di(2-ethylhexyl) terephthalate has been replacing phthalates at an increasing rate (Rodriguez-Carmona et al., 2019). PCPs for topical use frequently have UV filters or a UV-blocking ingredient in them. Sunscreen, cosmetics, and skin lotions all reported a greater quantity of benzophenone-3 (BP3), a substance that blocks ultraviolet light (Liao & Kannan, 2014). Thus, a tiny PCP packet exposes a person to a broad range of chemicals.

9.3 COMPOSITION AND USE OF PCPs

There aren't many datasets on how PCPs are used across various demographics. Men of today may be exposed to a broader variety of chemicals daily due to their propensity to be exposed to many PCPs at once. The overuse of PCPs highlights the necessity of considering the cumulative toxicological effects of concurrent exposure to many drugs (Dodson et al., 2012). Epidemiological studies are crucial for gathering the information required for toxicological risk assessment, consumer exposure assessments, or safety assessments. These kinds of activities require information on the ingredients and components of PCPs, their dosages, and the frequency of usage across different age groups (Loretz et al., 2005; Loretz et al., 2006; Hall et al., 2007). Because consumers and products have distinct activity patterns, PCP exposure levels vary greatly (Loretz et al., 2011). Consumers and products have distinct activity patterns, and PCP exposure levels vary greatly (Loretz et al., 2011).

Furthermore, the kind of PCPs, their usage, metabolism, dermal penetration, non-use patterns, and co-use all affect the computation of aggregate exposure to components (Cowan-Ellsberry & Robison, 2009). PCP use frequency is a very personal choice that is impacted by socioeconomic status and way of living. Factors including climate, education level, age, race, survey season, and gender have a substantial impact on the variability of PCP use patterns (Wu et al., 2011).

The amount and location of PCP use are influenced by the season, the nation, and the type of product. Income and occupation have also been shown to significantly alter how certain PCPs are used to the features that have already been mentioned (Lang et al., 2016; Park et al., 2018). Among other PCPs, lipstick, face cream, and body lotion are commonly used by American women. Different bodily parts may react negatively to many daily applications. Other PCPs that women frequently use include body wash, eye makeup, liquid foundation, face cleanser, perfume, and shampoo (Loretz et al., 2005, 2006, 2008).

Similar to this, households in California typically use 30 PCPs, with usage frequency varying based on the user's age and gender (Wu et al., 2011). It was also demonstrated that the PCP usage patterns varied according to the reproductive age of the females. For instance, the use of makeup and products used for hairstyling decreased during the postpartum period and pregnancy. Simultaneously, following delivery, more baby items are used, including diaper cream, baby wash, and baby lotion (Lang et al., 2016).

Moreover, it was discovered that women use lipstick more frequently than men do (Park et al., 2018). In a similar vein, women mostly take PCPs and prefer to use particular PCPs more frequently than men. Nonetheless, a comparable proportion of men and women were discovered utilizing special care items and liquid soaps

(Wu et al., 2011), excluding shaving products (Biesterbos et al., 2013). Similar to adult usage, children's PCP usage varies. Younger children mostly use body lotions and bath gel, while older children use antibacterial soap, hair mice, nail paint, hair conditioner, lip balm, and body lotion. Children who are older than 8 also use hair conditioner and deodorant.

However, it was revealed that girls used PCPs more than boys did. Notwithstanding, the fact that there were fewer sex differences in children's PCP usage (Wu et al., 2011). Ethnicity and race have an impact on the PCP use patterns frequency and prevalence. For example, there is diversity in the hair textures of African Americans. They use shampoo and conditioner less frequently than Asian women since they treat them permanently with chemical relaxants and straighteners.

Also, Asian women use skin care products more frequently (Wu et al., 2011). Age-based variations in PCP consumption patterns were also noted. It was seen that younger girls used PCPs at higher rates than older girls. Shampoo, conditioner, and hair mouse are utilized by young girls. Also, hair color, nail polish, mascara, foundation, hair spray, and some personal grooming goods were seen as more common among older females (Biesterbos et al., 2013; Wu et al., 2011). In South Korea, younger girls—those under 10 years old—were discovered using sun cream and spray more frequently than advanced girls (Oh et al., 2019).

In a similar vein, women's yearly hair coloring usage varies according to product kind and age (Bernard et al., 2016). According to research, the use of chemical relaxants and straighteners by females rises between infancy and adolescence and falls during maturity (Gaston et al., 2019). In contrast to older men who frequently use hair sprays and aftershave products, younger men prefer to use more of hair mousse and sunscreen frequently (Biesterbos et al., 2013; Wu et al., 2011).

Some PCPs have seasonal variations in their patterns of use; for example, lip balm was quite popular in California during summer (Wu et al., 2011). It was also discovered that the kind of PCP and gender affected how much product was used. Skincare goods like lotions, sunscreen, and creams are typically used extensively. Compared to men, women make use of more skincare, after-tanning, and suncare products. Similarly, women make use of more shaving products than men do, most likely because their areas of shaving are larger. The quantity of PCPs used was discovered to be influenced by age; for instance, younger and middle-aged women were more likely than older women to use eye shadow (Biesterbos et al., 2013). The PCP application area affects the level of exposure to many environmental pollutants as well. Since the skin permeability of anatomical sites affects the overall exposure to PCPs. The sequence of the scrotum determines a decrease in skin permeability: forehead extremities < back < scalp < axilla. Certain items, like sunscreen, body lotion, and skin care products, are applied over a bigger area of the body in a greater quantity because they are not limited to a particular body part. Legs, upper arms, and lower arms are the areas where over 90% of the users apply body creams. Men use shaving foam on their face and head, whereas women use it on their lower thighs, pelvic region, and axillae. Younger customers were discovered to use shaving foams more frequently than middle-aged and older consumers (Biesterbos et al., 2013). Additionally, some product types—like bath foam—are utilized at distinct times per day, for example, lotions, cosmetics, nail polish cleansers, etc., that are typically ingested at night or in the evening. Certain items, including lip balm and hand cream, can be used at any moment of the day (Biesterbos et al., 2013).

It was also found that a consumer's degree of education had an impact on how they used PCPs. Higher-educated individuals were observed to use more eye pencils, hair dye, and aftershave than intermediate or lower-educated individuals (Biesterbos et al., 2013; Wu et al., 2011). However, over 70% of users were discovered to use fragranced rather than non-fragranced PCPs, maybe as a result of the scarcity of non-fragranced products on the market. On the other hand, children used more non-fragranced hygiene products than adults (Wu et al., 2011). Additionally, Biesterbos and associates looked into the co-use patterns of 32 distinct PCPs. Women typically use 15 PCPs, whereas males use seven distinct PCPs, for a total of 13 PCPs used by males and females.

9.4 BIOCHEMICAL AND PHYSIOLOGICAL EFFECTS OF PCPs ON BACTERIA AND ALGAE COMMUNITY

Broad-spectrum antimicrobial compounds like triclocarban and triclosan are hazardous to freshwater crustaceans, algae, and tadpoles. They are discovered in numerous items like toothpaste, deodorant, and antimicrobial soaps. According to Heath and Rubin (1999), these substances work by preventing bacteria from synthesizing fatty acids. The same progression used by bacteria to synthesize fatty acids is also used by plants (Harwood, 1996).

Polyunsaturated fatty acids (PUFAs), which are essential to animal health as precursors of hormones and are important in cell membrane function, are almost exclusively found in plants (Brett & Muller-Navarra, 2003). The highly unsaturated fatty acids (HUFAs), a subset of polyunsaturated fatty acids (PUFAs), are important factors in assessing the suitability of algae as a dietary source (Brett & Muller-Navarra, 2003). Triclosan may deteriorate the attribute of algae that zooplankton uses as a food source. At environmentally relevant doses, triclosan has been seen to disrupt the production of fatty acids and decrease biomass and variety in algae (Wilson & Smith, 2003; Brain & Hanson, 2008).

The attribute of algal food is a critical component of the aquatic food web's health and functioning. The quality of the algae that zooplankton eats directly affects its growth and ability to withstand fish predation (Danielsdottir & Brett, 2007). Furthermore, when there is a healthy zooplankton community and excellent algal quality, production at higher trophic levels is improved through efficient energy transfer (Danielsdottir & Brett, 2007). Therefore, a reduction of algal species that can synthesize fatty acids and provide HUFAs to zooplankton due to disclosure to triclosan could result in adverse impacts on organisms throughout the food web, including fish species.

According to Franza et al. (2008), green, blue-green, and microalgae are the most sensitive to triclosan, with differences in sensitivity occurring up to two orders of magnitude. Environmentally relevant quantities of triclosan (0.015 µg/L to 2.8 µg/L) and triclocarban (10 µg/L to 30 µg/L) have been seen to cause chronic toxicity to algal communities (Wilson & Smith, 2003; Chalew & Halden, 2009). According to Chalew and Halden (2009), triclosan and triclocarban concentrations in the environment range from below detection limits to 2.3 μg/L and 0.25 μg/L, respectively. Triclocarban is, therefore, hazardous to algae at far higher concentrations than are normally encountered on water surfaces.

Concentrations of triclosan ranging from 6 μg/L to 182 μg/L and triclocarban from 0.06 μg/L to 4.7 μg/L have been seen to have negative effects on freshwater crustaceans (Chalew & Halden, 2009). Like algae, crustaceans play a significant role in the aquatic food chain, and exposure to triclocarban may have negative effects on the environment as a whole. At doses as low as 0.15 μg/L, pre-metamorphic North American bullfrog (*Rana catesbeiana*) tadpoles showed signs of endocrine disruption by triclosan through interference with thyroid hormones, which regulates metamorphosis of tadpoles into froglets (Veldhoen & Skirrow, 2006). When disclosed to triclosan at 2.3 μg/L, the tadpoles of the African clawed frog (*Xenopus laevis*) showed decreased activity, whereas the tadpoles of the American toad (*Bufo americanus*) and northern leopard frog (*Rana pipiens*) showed earlier and higher mortality (Fraker & Smith, 2004; Smith & Burgett, 2005).

According to the studies mentioned earlier, at environmentally relevant doses, triclosan and triclocarban are hazardous to freshwater crustaceans, algae, and frog species through at least two different pathways. Having been examined more than triclocarban, triclosan has more information available. However, it is anticipated that these substances will severely reduce the attributes of the environment because they have been seen to be hazardous to aquatic creatures over an extended period.

9.5 EFFECTS OF PCPs ON AQUATIC ECOSYSTEMS

The impact that PCPs on water surfaces have on aquatic creatures and ecosystems is still largely unknown. Even though acute toxicity has been the subject of numerous research, PCPs are rarely encountered in the environment in sufficient quantities to trigger an acute response; therefore, this is usually not a cause for concern (Halling-Sorensen & Nielsen, 1998). Over the course of several generations, exposure to the aquatic environment is continuous (Daughton & Ternes, 1999; Fent & Weston, 2006). Understanding the long-term impacts is essential to comprehend the potential consequences of PCPs on aquatic organisms because of their "pseudo-persistent" nature and low concentrations. Understanding the effects on aquatic organisms involves comprehending how environmental factors, such as pH, affect toxicity. By design, PCPs are biologically active compounds, and while they are designed for target organism physiology, many physiological pathways and receptor targets are evolutionarily conserved (Beulig & Fowler, 2008; Brain & Hanson, 2008). A toxicological response to a bioactive compound in an aquatic organism can result when the organism uses the same receptor or pathway as in humans with enzymes that are almost structurally identical (Brain & Hanson, 2008).

Through hormone-mimicking, blocking, or disruption, certain PCPs function as endocrine-disrupting chemicals or compounds (EDCs) in the environment, interfering with an organism's natural hormonal function. Estrogenic, androgenic, and thyroidal chemicals are the three main categories of EDCs (Snyder & Westerhoff, 2003). There isn't a full list of PCPs or EDCs that meet the requirements to be considered EDCs (Kim & Cho, 2007). Bioaccumulation and the biochemical reaction in non-target organisms can be facilitated by properties like persistence, which prevents inactivity before the therapeutic effect is reached, and lipophilicity, which enables transport across membranes (Halling-Sorensen & Nielsen, 1998). Furthermore, pharmacodynamics may be detrimental to aquatic creatures in addition to the intended therapeutic impact (Fent & Weston, 2006). Although research on long-term impacts has been done, the effects of PCPs on freshwater aquatic environments are not well understood (Fent & Weston, 2006; Kummerer, 2009).

Furthermore, PCPs have been implicated in numerous published research on chronic effects at concentrations far greater than those found in the environment (Halling-Sorensen, 2000). Studies that used doses significantly higher than those found in the environment frequently showed that PCPs are harmful to aquatic life. Nevertheless, the results of those experiments are less helpful in determining the effects of PCPs on aquatic creatures because it is unlikely that PCPs in such high concentrations will be encountered. The chronic effects of PCPs on aquatic animals and ecosystems that are known to occur at environmentally relevant concentrations are discussed later, along with the physiological pathways and receptors that are targeted when relevant data is available.

9.6 PERSISTENCE DEGRADATION OF PCPs POLLUTANTS IN WATER BODIES

Long-term exposure to aquatic life may occur from PCPs' propensity to linger in aquatic habitats. PCP persistence in aquatic settings can be impacted by several factors, including biological processes, ambient conditions, and the physicochemical characteristics of the compounds (Wang et al., 2022).

9.7 PERSISTENCE DEGRADATION OF PCPs POLLUTANTS IN WATER BODIES

Long-term exposure to aquatic life may occur from PCPs' propensity to linger in aquatic habitats. PCP persistence in aquatic settings can be impacted by several factors, including biological processes, ambient conditions, and the physicochemical characteristics of the compounds (Wang et al., 2022). The following are some instances of PCP persistence in aquatic life.

9.7.1 Hormone-Related Substances

PCPs frequently contain hormonal chemicals like β-estradiol and estrone, which have been demonstrated to linger for a long time in aquatic habitats. These substances can cause reproductive disorders in aquatic life by interfering with endocrine function (Fang et al., 2019).

9.7.2 Personal Care Products

Personal care products, such as sunscreen and insect repellent, can persist in aquatic environments for extended periods, leading to potential harm to aquatic life. For example, the compound oxybenzone, which is commonly found in sunscreen, is toxic to coral reefs and other marine life (Xiang et al., 2021).

9.7.3 Antidepressants

Antidepressants, such as fluoxetine and sertraline, are commonly detected in aquatic environments and can persist for extended periods. These compounds can affect the behavior of aquatic organisms and lead to changes in their feeding, reproduction, and other activities (Chaturvedi et al., 2021).

Concern should be expressed about PCPs' persistence in aquatic habitats since it may result in long-term exposure and possible harm to aquatic life. To lessen PCP's negative environmental effects and safeguard aquatic ecosystems, effective management techniques are required, such as enhanced wastewater treatment and PCP usage regulations (Chaturvedi et al., 2021).

9.8 BIOACCUMULATION AND PHYSIOLOGICAL EFFECTS OF PCPs ON BACTERIA AND ALGAE COMMUNITY

Because PCPs can bioaccumulate over time in the tissues of aquatic species, they can be a serious threat to these creatures. When an organism absorbs a chemical more quickly than it can expel it, a buildup of the chemical happens in the organism's tissues, a process known as bioaccumulation (Keerthanan et al., 2022). Some examples of surface water microorganisms and the effects of personal care products on them are shown in Table 9.1 later.

The following are some causes and dangers linked to PCP bioaccumulation in aquatic animals.

9.8.1 Slow Elimination

In aquatic species, PCPs can have extended elimination half-lives, which can eventually cause a buildup of these substances in the tissues of the organisms. Potential toxicity and injury to the organisms may result from this.

9.8.2 Low Solubility

A lot of PCPs bind to organic materials and are not very soluble in water, which increases the likelihood that they will build up.

9.8.3 Trophic Transfer

PCPs can build up in greater quantities in the tissues of predators that eat infected prey, which raises the concentration of these substances further up the food chain.

9.8.4 Organism-Specific Factors

Depending on the species, sex, age, and other biological characteristics of the organism, the rate of bioaccumulation can change. This may result in varying PCP buildup

TABLE 9.1
Effects of Personal Care Products on Some Surface Water Microorganisms.

Surface water microorganisms	Effects of personal care products
Algae	Personal care products containing phosphates, nitrates, or other nutrients can lead to excessive algal growth, causing algal blooms and disrupting the natural balance of the ecosystem. Some ingredients in personal care products, such as certain surfactants, can also have toxic effects on algae.
Bacteria	Personal care products that contain antimicrobial substances like triclosan or triclocarban can inhibit the growth of bacteria in surface water. However, this can also lead to the development of antimicrobial-resistant bacteria over time, which poses a problem for the ecosystem.
Protozoa	Personal care products can potentially impact protozoa in surface water. Some ingredients like surfactants can disrupt the cell membranes of protozoa, leading to their death. In addition, antimicrobial substances can also affect the growth and survival of certain species.
Fungi	Some personal care products such as antifungal creams or shampoos may contain substances that specifically target fungi. If these products are improperly disposed of or enter surface water, they can potentially affect fungal populations, inhibiting their growth or even causing fungi to die.

in various aquatic animal populations. The organisms and the habitats they live in may be threatened by the PCPs that aquatic animals bioaccumulate. These dangers may be toxic. The accumulation of PCPs in aquatic animal tissues can lead to impaired growth and development, which can affect the overall health and survival of the organisms.

9.9 CONCLUSION

This chapter reviewed the effects of personal care products on surface water microorganisms. PCPs have the potential to be harmful to aquatic life, especially if they bioaccumulate over time in high concentrations. Reproductive problems, neurological diseases, and other unfavorable health outcomes are examples of these effects. Therefore, PCPs linkages with the aquatic environment should be avoided, and future research should be carried out on PCPs to ascertain the actual concentrations of the PCPs that adversely affect the aquatic environment and its components.

9.10 RECOMMENDATION

To mitigate the potential negative effects of personal care products on surface water microorganisms, it is essential to use and dispose of these products responsibly. Choosing products that are eco-friendly, biodegradable, and free from harmful chemicals can help minimize the impact on aquatic ecosystems. Additionally, proper wastewater treatment and disposal practices can help reducc the release of these substances into surface waters.

REFERENCES

Allmyr M., Adolfsson-Erici M., McLachlan M.S. and Sandborgh-Englund G., 2006. Triclosanin plasma and milk from Swedish nursing mothers and their exposure via personal care products. *Sci. Total Environ.* 372: 87–93.

Bernard A., Houssin A., Ficheux A.S., Wesolek N., Nedelec A.S., Bourgeois P., Hornez N., Batardière A., Misery L. and Roudot A.C., 2016 Consumption of hair dye products by the French women population: Usage pattern and exposure assessment. *Food Chem. Toxicol.* 88: 123–132.

Beulig A. and Fowler J., 2008. Fish on prozac: Effect of serotonin reuptake inhibitors on cognition in goldfish (vol 122, pg 426, 2008). *Behav. Neurosci.* 122(4): 776–776.

Bickers D.R., Calow P., Greim H.A., Hanifin J.M., Rogers A.E., Saurat J.H., Glenn Sipes I., Smith R.L. and Tagami H., 2003. The safety assessment of fragrance materials. *Regul. Toxicol. Pharmacol.* 37: 218–273.

Biesterbos J.W.H., Dudzina T., Delmaar C.J.E., Bakker M.I., Russel F.G.M., Von Goetz N., Scheepers P.T.J. and Roeleveld N., 2013. Usage patterns of personal care products: Important factors for exposure assessment. *Food Chem. Toxicol.* 55: 8–17.

Brain R.A. and Hanson M.L., 2008. Aquatic plants exposed to pharmaceuticals: Effects and risks. *Rev. Environ. Contam. Toxicol.* 192: 67–115.

Brett M.T. and Muller-Navarra D.C., 2003. The role of highly unsaturated fatty acids in aquatic food web processes. *Freshwater Biology*, 38(3): 483–499. First published: 18. https://doi.org/10.1 046/j.1365-2427.1997.00220.x

Chalew T.E.A. and Halden R.U., 2009. Environmental exposure of aquatic and terrestrial biota to triclosan and triclocarban. *J. Am. Water Resour. Assoc.* 45(1): 4–13.

Chaturvedi P., Shukla P., Giri B.S., Chowdhary P., Chandra R., Gupta P. and Pandey A., 2021. Prevalence and hazardous impact of pharmaceutical and personal care products and antibiotics in environment: A review on emerging contaminants. *Environ. Res.* 194: 110664.

Collado N., Rodriguez-Mozaz S., Gros M., Rubirola A., Barceló D., Comas J., Rodriguez-Roda I. and Buttiglieri G., 2014. Pharmaceuticals occurrence in a WWTP with significant industrial contribution and its input into the river system. *Environ. Pollut.* 185: 202–212.

Costanzo S.D., Watkinson A.J., Murby E.J., Kolpin D.W. and Sandstrom M.W., 2007. Is there a risk associated with the insect repellant DEET (N,N-diethyl-mtoluamide) commonly found in aquatic environments? *Sci. Total Environ.* 384: 214–220.

Cowan-Ellsberry C.E. and Robison S.H., 2009. Refining aggregate exposure: Example using parabens. *Regul. Toxicol. Pharmacol.* 55: 321–329.

Danielsdottir M.G. and Brett M.T., 2007. Phytoplankton food quality control of planktonic food web processes. *Hydrobiologia.* 589: 29–41.

Daughton C.G. and Ternes T.A., 1999. Pharmaceuticals and personal care products in the environment: Agents of subtle change? *Environ. Health Perspect.* 107: 907–938.

Dobbins L.L., Usenko S., Brain R.A., Brooks B.W., 2009. Probabilistic ecological hazard assessment of parabens using Daphnia magna and Pimephales promelas. *Environ. Toxicol. Chem.* 28: 2744–2753.

Dodson R.E., Nishioka M., Standley L.J., Perovich L.J., Brody J.G. and Rudel R.A., 2012. Endocrine disruptors and asthma-associated chemicals in consumer products. *Environ. Health Perspect.* 120: 935–943.

Fang W., Peng Y., Muir D., Lin J. and Zhang X., 2019. A critical review of synthetic chemicals in surface waters of the US, the EU and China. *Environ. Int.* 131: 104994.

Fenech C., Nolan K., Rock L. and Morrissey A., 2013. An SPE LC-MS/MS method for the analysis of human and veterinary chemical markers within surface waters: An environmental forensics application. *Environ. Pollut.* 181: 250–256.

Fent K. and Weston A.A., 2006. Ecotoxicology of human pharmaceuticals. *Aquat. Toxicol.* 76(2): 122–159.

Fiume M.M., Heldreth B., Bergfeld W.F., Belsito D.V., Hill R.A., Klaassen C.D., Liebler D.C., Marks J.G., Shank R.C., Slaga T.J., Snyder P.W. and Andersen F.A., 2017. Safety assessment of diethanolamine and its salts as used in cosmetics. *Int. J. Toxicol.* 36: 89S–110S.

Fraker S.L. and Smith G.R., 2004. Direct and interactive effects of ecologically relevant concentrations of organic wastewater contaminants on Ranapipiens tadpoles. *Environ. Toxicol.* 19(3): 250–256.

Franza S., Altenburger R., Heilmeierb H. and Schmitt-Jansena M., 2008. What contributes to the sensitivity of microalgae to triclosan? *Aquat. Toxicol.* 90: 102–108.

Frédéric O. and Yves P., 2014. Pharmaceuticals in hospital wastewater: Their ecotoxicity and contribution to the environmental hazard of the effluent. *Chemosphere.* 115(1): 31–39.

Gaston S.A., James-Todd T., Harmon Q., Taylor K.W., Baird D. and Jackson C.L., 2019. Chemical/straightening and other hair product usage during childhood, adolescence, and adulthood among African-American women: Potential implications for health. *J. Expo. Sci. Environ. Epidemiol.* 30: 86–96.

Hall B., Tozer S., Safford B., Coroama M., Steiling W., Leneveu-Duchemin M.C., McNamara C. and Gibney M., 2007. European consumer exposure to cosmetic products, a framework for conducting population exposure assessments. *Food Chem. Toxicol.* 45: 2097–108.

Halling-Sorensen B. (2000). Algal toxicity of antibacterial agents used in intensive farming. *Chemosphere.* 40(7): 731–739.

Halling-Sorensen B. and Nielsen S.N., 1998. Occurrence, fate and effects of pharmaceutical substances in the environment—A review. *Chemosphere.* 36(2): 357–394.

Harwood J.L., 1996. Recent advances in the biosynthesis of plant fatty acids. *Biochim. Biophys. Acta Lipids Lipid Metab.* 1301(1–2): 7–56.

Heath R.J. and Rubin J.R., 1999. Mechanism of triclosan inhibition of bacterial fatty acid synthesis. *J. Biol. Chem.* 274(16): 11110–11114.

Iglesias A., Nebot C., Vázquez B.I., Miranda J.M., Abuín C.M.F. and Cepeda A., 2014. Detection of veterinary drug residues in surface waters collected nearby farming areas in Galicia, North of Spain. *Environ. Sci. Pollut. Res.* 21(3): 2367–2377.

Iyer M., Tiwari S., Renu K., Pasha M.Y., Pandit S., Singh B., Raj N., Krothapalli S., Kwak H.J., Balasubramanian V. and Jang S.B., 2021. Environmental survival of SARS-CoV-2—a solid waste perspective. *Environ. Res.* 197: 111015.

Keerthanan S., Jayasinghe C., Biswas J.K., Vithanage M., 2022. Pharmaceutical and Personal Care Products (PPCPs) in the environment: Plant uptake, translocation, bioaccumulation, and human health risks. *Crit. Rev. Environ. Sci. Technol.* 51(12): 1221–1258.

Kim S.D. and Cho J., 2007. Occurrence and removal of pharmaceuticals and endocrine disruptors in South Korean surface, drinking, and waste waters. *Water Res.* 41(5): 1013–1021.

Kolpin D.W., Furlong E.T., Meyer M.T., Thurman E.M., Zaugg S.D., Barber L.B. and Buxton H.T., 2002. Pharmaceuticals, hormones, and other organic wastewater contaminants in U.S. streams, 1999–2000: A national reconnaissance. *Environ. Sci. Technol.* 36(6): 1202–1211.

Kummerer K., 2009. Antibiotics in the aquatic environment—A review—Part II. *Chemosphere.* 75(4): 435–441.

Lang C., Fisher M., Neisa A., MacKinnon L., Kuchta S., MacPherson S., Probert A. and Arbuckle T., 2016. Personal care product use in pregnancy and the postpartum period: Implications for exposure assessment. *Int. J. Environ. Res. Public Health.* 13: 105.

Liao C. and Kannan K., 2014. Widespread occurrence of benzophenone-type UV light filters in personal care products from China and the United States: An assessment of human exposure. *Environ. Sci. Technol.* 48: 4103–4109.

Loretz C., Von Goetz N., Scheringer M., Wormuth M. and Hungerbühler K., 2011. Potential exposure of German consumers to engineered nanoparticles in cosmetics and personal care products. *Nanotoxicology.* 5: 12–29.

Loretz L.J., Api A.M., Babcock L., Barraj L.M., Burdick J., Cater K.C., Jarrett G., Mann S., Pan Y.H.L., Re T.A., Renskers K.J. and Scrafford C.G., 2008. Exposure data for cosmetic products: Facial cleanser, hair conditioner, and eye shadow. *Food Chem. Toxicol.* 46: 1516–1524.

Loretz L.J., Api A.M., Barraj L.M., Burdick J., Davis D.A., Dressler W., Gilberti E., Jarrett G., Mann S., Laurie Pan Y.H., Re T., Renskers K., Scrafford C. and Vater S., 2006. Exposure data for personal care products: Hairspray, spray perfume, liquid foundation, shampoo, body wash, and solid antiperspirant. *Food Chem. Toxicol.* 44: 2008–2018.

Loretz L.J., Api A.M., Barraj L.M., Burdick J., Dressler W.E., Gettings S.D., Han Hsu H., Pan Y.H.L., Re T.A., Renskers K.J., Rothenstein A., Scrafford C.G. and Sewall C., 2005 Exposure data for cosmetic products: Lipstick, body lotion, and face cream. *Food Chem. Toxicol.* 43: 279–291.

Lu S., Yu Y., Ren L., Zhang X., Liu G. and Yu Y., 2018. Estimation of intake and uptake of bis phenols and triclosan from personal care products by dermal contact. *Sci. Total Environ.* 621: 1389–1396.

Oh M., Kim S., Han J., Park S., Kim G.U. and An S., 2019. Study on consumer exposure to sun spray and sun cream in South Korea. *Toxicol. Res.* 35: 389–394.

Orii Y. and Kannan K., 2008. Survey of organosilicone compounds, including cyclic and linear siloxanes, in personal-care and household products. *Arch. Environ. Contam. Toxicol.* 55: 701.

Park G.H., Nam C., Hong S., Park B., Kim H., Lee T., Kim K., Lee J.H. and Kim M.H., 2018. Socioeconomic factors influencing cosmetic usage patterns. *J. Expo. Sci. Environ. Epidemiol.* 28: 242–250.

Patel R.V., Shaeer K.M., Patel P., Garmaza A., Wiangkham K., Franks R.B., Pane O. and Carris N.W., 2016. EPA-Registered repellents for mosquitoes transmitting emerging viral disease. *Pharmacotherapy.* 36: 1272–1280.

Reiner J.L. and Kannan K.A., 2006. Survey of polycyclic musks in selected household commodities from the United States. *Chemosphere.* 62: 867–873.

Rodríguez-Carmona Y., Ashrap P., Calafat A.M., Ye X., Rosario Z., Bedrosian L.D., Huerta-Montanez G., Vélez-Vega C.M., Alshawabkeh A. and Cordero J.F., 2019. Determinants and characterization of exposure to phthalates, DEHTP and DINCH among pregnant women in the Protect birth cohort in Puerto Rico. *J. Expo. Sci. Environ. Epidemiol.* 1–14.

Smith G.R. and Burgett A.A., 2005. Effects of three organic wastewater contaminants on American toad, Bufoamericanus, tadpoles. *Ecotoxicology.* 14(4): 477–482.

Snyder S.A. and Westerhoff P., 2003. Pharmaceuticals, personal care products, and endocrine disruptors in water: Implications for the water industry. *Environ. Eng. Sci.* 20(5): 449–469.

Soni M.G., Burdock G.A., Taylor S.L. and Greenberg N.A., 2001. Safety assessment of propyl paraben: A review of the published literature. *Food Chem. Toxicol.* 39: 513–532.

Veldhoen N. and Skirrow R.C., 2006. The bactericidal agent triclosan modulates thyroid hormone-associated gene expression and disrupts postembryonic anuran development. *Aquat. Toxicol.* 80(3): 217–227.

Wang C., Lu Y., Wang C., Xiu C., Cao X., Zhang M. and Song S., 2022. Distribution and ecological risks of pharmaceuticals and personal care products with different anthropogenic stresses in a coastal watershed of China. *Chemosphere.* 303: 135176.

Wang R., Moody R.P., Koniecki D. and Zhu J., 2009. Low molecular weight cyclic volatile methylsiloxanes in cosmetic products sold in Canada: Implication for dermal exposure. *Environ. Int.* 35: 900–904.

Wilson B.A. and Smith V.H., 2003. Effects of three pharmaceutical and personal care products on natural freshwater algal assemblages. *Environ. Sci. Technol.* 37(9): 1713–1719.

Wu X., Bennett D.H., Ritz B., Cassady D.L., Lee K. and Hertz-Picciotto I., 2011. Usage pattern of personal care products in California households. *Food Chem. Toxicol.* 48: 3109–3119.

Xiang Y., Wu H., Li L., Ren M., Qie H. and Lin A., 2021. A review of distribution and risk of pharmaceuticals and personal care products in the aquatic environment in China. *Ecotoxicol. Environ. Saf.* 213: 112044.

10 Physical and Biological Removal of the Mass Load of Emergent Pollutants from Waste Treatment Facilities

Linda U. Obi, Frances N. Olisaka, Felix C. Onyia, Israel H. Innocent, and Paschaleen Onyemaechi

10.1 INTRODUCTION

One of the major effects of climate change is the difficulty in accessing water, and a broad and joint approach must be used to manage this limited resource if we are to increase resilience against climate change and provide for a growing population. According to the United Nations (UN) 2030 Agenda for Sustainable Development, Sustainable Development Goal (SDG) 6, which focuses on water and sanitation, also incorporates issues related to the management of water, wastewater, water resources, and ecosystems (Patel et al., 2020).

Contaminants of emerging concern (CECs) are also referred to as emergent pollutants (EPs); they are compounds that are introduced into the environment but for which there are currently no recognized restrictions (Wiest et al., 2021). Wiest et al. (2021) also explained that due to the increase in the global population and socioeconomic development in various sectors, such as industries, agriculture, and health, coupled with the prevailing changes in climatic conditions, increased water consumption has led to the generation of significant quantities of waste. Urban wastewater and effluent discharged from wastewater treatment facilities represent prominent sources of most emergent pollutants in aquatic ecosystems, and these EPs are toxic and typically persist in the environment, which makes them potentially dangerous to ecosystems and to the general public's health (Wiest et al., 2021). Emergent pollutants are naturally occurring or of synthetic origin, and they include pesticides, hormones, microplastics, personal care products, pharmaceuticals, and surfactants (Mahmood et al., 2022). The need for their removal is essential since these wastes are persistent in the environment and can bioaccumulate. To manage the accruing waste and mitigate the challenges of potential contamination of the surrounding

DOI: 10.1201/9781003362975-10

ecosystem and destruction of habitats for wildlife, wastewater treatment facilities are constructed to treat and remove these wastes.

The treatment or removal of EP waste from treatment facilities involves four (4) distinct stages/levels, including preliminary, primary, secondary, and tertiary stages as described by Mahmood et al. (2022). The preliminary and primary phases are the initial phases that deal with screening and basic removal of physical wastes while the secondary phase is the implementation of techniques that will remove EPs that escaped the primary treatment phase. The tertiary treatment level removes about 85%–95% of suspended solids and 50%–90% of dissolved solids. While some municipal wastewater treatment facilities use a tertiary stage of treatment for EPs, the majority use primary and secondary levels of treatment (Mahmood et al., 2022).

Waste treatment facilities utilize some conventional technologies in their management of EPs resident in wastewater, and these include physical, chemical, and biological technologies. The use of just one technique for wastewater treatment is typically insufficient to provide the best results (Srivastava et al., 2021). According to Srivastava et al. (2021), Combining multiple procedures and performing them in sequence could have a favorable outcome. However, environmental sustainability depends on waste removal techniques that use both physical and biological methods. The physical techniques focus on straightforward design and cost-effectiveness. However, the selection of a physical removal process is based on the type of waste and its properties. Although cost-effective and efficient, biological techniques are inefficient for some EPs such as synthetic dyes; this is due to the ability of such EPs to resist aerobic biodegradation (Srivastava et al., 2021).

Nevertheless, chemical treatments typically have toxic by-products and are ineffective (Mahmood et al., 2022). Physical strategies for the removal of EPs from treatment facilities include sedimentation, filtration, stabilization/solidification, floatation, coagulation, etc. Sedimentation is a physical water treatment strategy that employs the force of gravity in the removal of suspended materials from water. A subsequent waste removal technique, filtration removes solid particles from a liquid or gaseous fluid by passing it through a filter membrane (Shen et al., 2021). The biological waste removal strategies employ the degradation potentials of microorganisms, and they encompass activated sludge, trickling filters, etc.

This chapter focuses on the stages of EP treatment as well as the techniques used to physically and biologically remove EPs from waste facilities. The potential challenges and prospects for the effective removal of EPs from waste treatment plants were also addressed.

10.2 ORIGIN AND TYPES OF EMERGING POLLUTANTS

Emerging pollutants are usually introduced into the environment through human activities. These emerging pollutants (EP) are made up of substances such as pharmaceuticals, personal care products (PCPs), surfactants, plasticizers, pesticides, industrial chemicals, additives, endocrine disruptors, microbeads, etc. (Patel et al., 2020).

10.2.1 Pharmaceuticals

A variety of chemical substances have been used to formulate certain products to meet human satisfaction. These products include cosmetics, pharmaceuticals, personal care products, pesticides, insect repellents, synthetic hormones, artificial sweeteners, as well as cleansing products (Kolpin et al., 2002).

Pharmaceuticals are chemical compounds that are either naturally secreted or artificially blended with the ultimate intention of preventing the spread of disease, curing the disease, or enhancing human life (Maletz et al., 2013). Different compounds associated with these pharmaceutical products have been detected in both fresh and wastewaters worldwide (Vulliet and Cren-Olive, 2011). Based on therapeutic use, pharmaceutical drugs in circulation include antibiotics (ciprofloxacin), antimicrobials (penicillin), anti-epileptic (carbamazepine), anti-diabetics, anti-inflammatories, analgesics, anti-histamines, anti-ulcer, anti-anxiety (diazepam), and lipid regulators (clofibrate) (Kanakaraju et al., 2014). The main classes of pharmaceutical drugs that are prevalently available in water bodies include nonsteroidal anti-inflammatory drugs, anticonvulsants, antibiotics, and lipid regulators (Mompelat et al., 2009). Pharmaceutical products are compromised by their chronically toxic effects because their compounds are biologically undegradable and water-insoluble, thus are available in both waste and freshwater bodies.

10.2.2 Personal Care Products (PCPs)

These are products produced from different substances, both active and inert. These substances include both prescribed and non-prescribed pharmaceuticals which are used by humans for personal care purposes and animals for veterinary purposes (Jiang et al., 2013). These products include synthetic hormones, cosmetics, preservatives, toiletries, fragrances, insect repellent, shampoos, lipid regulators, steroids, as well as analgesics as described by Jiang et al. (2013). Most of these products are routinely used and, thus, common to us. However, they are not ingested like pharmaceutical products; rather, they are used topically (Jiang et al., 2013). Personal care products can be grouped into two, these include polycyclic musks as well as parabens (Fawell and Ong, 2012). Many consumer products, such as hand soaps, air fresheners, lotions, toothpaste, etc., are manufactured with disinfectants like chloroprene and triclosan (Fawell and Ong, 2012).

Recently, there has been the identification of personal care products (PCPs) (in their original or transformed state) from water bodies, including water treatment plants due to their incessant usage by the populace. They end up in the community water system and soil after being introduced from leached farmland manure/sewage treatment plants, swimming, bathing, and showering (Parra-Saldivar et al., 2020). These PCPs are fully converted into harmless substances such as carbon dioxide and water, while the lipophilic and non-biodegradable products get attached to other hydrophilic compounds as they get absorbed partially into sedimentation sludge (Jiang et al., 2013). The more abundant substances in the environment are the quantity of the PCPs that are introduced either in their originality or transformed/metabolized state in the wastewater treatment plants.

10.2.3 Endocrine Disruptive Chemicals (EDCs)

Certain compounds have been identified to interrupt the function of hormones secreted by the endocrine gland. These compounds are referred to as 'endocrine-disrupting compounds/modulators' (Ferguson et al., 2013). These compounds are available in products such as cosmetics, plastic bottles, detergents, children's toys, etc. These have been documented to have caused harmful effects on the endocrine systems of aquatic life including fish, and these chemicals modify the body physiology as well as trigger the basic regulatory function in the body fluid (Ferguson et al., 2013). These effects may be transgenerational as well as irreversible (Patel et al., 2020). These compounds could be introduced into the environment both synthetically (through organic chemicals from some anthropogenic activities like bisphenol A) and naturally (through estrones, androgens, gestagens) (Patel et al., 2020). In a study by Tremblay et al. (2018), it was discovered that the basic source of steroid estrogens is crop residues and that estrogens are present in soil drainages, streams, stock grazing fields, runoff from dairy wastes, groundwater, and lands

10.2.4 Antiseptics

Antiseptics such as triclosan are known to be present in household appliances, toothpaste, soaps, clothing, fabrics, etc. When present in tap water, they yield chloroform as a chlorinated by-product (Gwenzi et al., 2018). Triclosan is broken down into dioxins in the environment. These include 2,4-dichlorophenol and 2,8-dichlorodibenzo-p-dioxin. Triclosan has been discovered in many water bodies in the United States, and its complete removal has not been achieved even with the conventional treatment methods; thus, its safety standards in drinking water are yet to be established (Gwenzi et al., 2018).

10.2.5 Perfluorinated Compounds (PFCs)

These compounds are found in paints, polishes, waxes, electronics, food packages, etc. Examples include perfluorocarbon sulfonic acids (PFSAs), perfluorinated carboxylic acids (PFCAs), perfluoro octane sulfonate (PFOS), and perfluorooctanoic acid (PFOA), while the last two are most commonly in use (Bai et al., 2018). There is detection of PFOA in drinking and treated water sources, possibly due to their release from industrial facilities and non-point sources (Bai et al., 2018). Their existence has also been observed in drinking water wells through contaminated groundwater. PFOA can gain access to groundwater through the atmosphere from close industrial sites up to 20 miles away before being deposited onto soil and seepage into underground water (Zhu and Kannan, 2019).

10.2.6 Disinfection By-products (DBPs)

Drinking water treatment plants have been observed to be the basic source of disinfection by-products (Stalter et al., 2020). Halogenic acetic acids (HAAs) and trihalomethane (THMs) were discovered to account for 38.1% and 42.6% of all DBPs, respectively, in all treated samples. Depending on the water source, the level of DBPs in drinking water has been observed to vary with surface water having the highest levels, mixed water sources, and groundwater having the lowest levels (Stalter et al., 2020).

10.2.7 Pesticides

The possible means of introducing pesticides into drinking water is a result of carelessness such as sizeable accidental spills, siphoning, and application in lawns (Srivastav et al., 2020). The metabolites of pesticides have been observed to be heavily released in the soil through leaching, and most of these metabolites in water matrices showed prevalent wide usage in city centers (Gallé et al., 2020).

10.3 POSSIBLE EFFECTS OF EMERGING POLLUTANTS ON THE ECOSYSTEM

Adverse effects of emerging pollutants have become a serious concern to the entire populace because of being implicated in causing cancer and disruption of the endocrine system. Endocrine disruption chemicals negatively affect the hormones of living organisms (Jiang et al., 2013). The normal hormone level could be negatively affected by this disruption, thereby resulting in abnormal regulation of hormone secretion and metabolism as stated by Afshan et al. (2020). The sexual development of a mother's offspring can be negatively affected when she gets exposed to any of these endocrine-disrupting compounds such as bisphenol A and alkylphenol ethoxylate, (Afshan et al., 2020). Secondary metabolites are generated when the pharmaceutical and personal care products (parent compounds) present in our environment are biotransformed, and the impact of these metabolites could be more toxic than the parent compounds with an unclear adverse effect on non-target organisms (WHO, 2015). Prolonged exposure of living organisms (including aquatic life) to these compounds endangers their lives heavily as a result of acute toxicity.

The possible ecological threats posed by antibiotics in marine ecosystems cannot be over-emphasized.

Lin et al. (2024) investigated the impact of two antibiotics, levofloxacin (LEV) and norfloxacin (NOR), on the cell growth parameter and eventual physiological variation of the marine microalgae *Skeletonema costatum* after exposure. It was observed that both drugs repressed the growth of *S. costatum* at all concentrations except at 10 mg/L for NOR. Another study by Pradit et al. (2024) discovered the abundance of heavy metals and microplastics in lagoon sediments. This may serve as a source of toxicity to marine organisms and humans, as it will greatly affect the food chain of the marine habitat.

10.4 DIFFERENT PHASES INVOLVED IN THE TREATMENT OF EMERGENT POLLUTANTS

10.4.1 Preliminary Phase of Treatment of Emergent Pollutants

The removal of large organic and inorganic debris present in flowing water, such as huge wood, boulders, plastics, dead animals, and tree branches, is part of the preliminary treatment of emerging contaminants in wastewater (Oakley and Von Sperling, 2017). It is the first treatment operation used in the removal of emerging pollutants in wastewater and has no direct benefit in terms of improving water purity; rather, its main importance is in protecting the water treatment equipment and infrastructure from clogging while also ensuring their long-term viability. Over time, the accumulation of debris in the treatment facility can erode it and cause wear and tear (Metcalf and Eddy, 2014). This phase consists of two major processes.

10.4.1.1 Screening

The screening procedure involves removing huge floating or suspended material from moving waters. If not removed, this debris could clog the pipes and inhibit efficient water treatment in the primary, secondary, and tertiary phases of treatment (Oakley and Von Sperling, 2017). Screens for this treatment range from coarse to fine, and they are often made of steel set parallel to one another or iron bars with small apertures of approximately an inch. Mesh screens with fewer apertures could be used to make other varieties. They are normally installed in chambers and face the direction of the flowing water. The debris is trapped at the screen's surface and can be cleaned both manually and mechanically (Oakley and Von Sperling, 2017). For large-scale wastewater treatment plants, a device called a comminutor or a barminutor is utilized in large-scale treatment plants. These devices capture and chop huge and heavy garbage to make collection easier (Oakley and Von Sperling, 2017). After being filtered, the wastewater is routed into the grit chamber for further treatment.

10.4.1.2 Grit Removal

Grits are heavy inorganic solids with specific gravities ranging from 2 mm to 2.65 mm, such as sands, gravel, heavy sticks, and metal bits. Allowing the water to settle causes tiny particles with a higher mass compared to the particles surrounding them to be gravity-sorted, or drop to the bottom of the mixture (Lee et al., 2020). Here, grit chambers are used to collect particles and transport them to a precise spot where they may be securely and effectively removed. Grit removal will be accomplished by automated gear in larger water treatment plants, although smaller plants may do so manually. Grit collection over time may cause excessive wear and abrasion at various treatment phases and must consequently be eliminated by utilizing grit chambers. After collecting these grits, they are either buried or utilized as manure (Oakley and Von Sperling, 2017)

10.4.2 Primary Phase of Treatment of Emergent Pollutants

The primary step of emergent pollutant treatment comprises filtration and sedimentation to remove a percentage of the suspended solids (scum) and organic matter from wastewater to decrease debris for further downstream treatment procedures (Metcalf and Eddy, 2014). It is sometimes referred to as primary sedimentation. The primary treatment's goal is to settle material by gravity, eliminate floatable debris, and minimize pollutants to make later treatment easier. The primary goal of wastewater treatment is to minimize the biochemical oxygen demand (BOD) and total suspended solids (TSS). BOD refers to the amount of dissolved oxygen consumed by aerobic microorganisms when organic matter is decomposed in water, while TSS refers to suspended particles in wastewater. The sediment, known as primary sludge, is highly decomposable and must be cleaned from the bottom of the sedimentation tank that holds the wastewater regularly (Oakley and Von Sperling, 2017). This is generally accomplished by anaerobic digestion as described in Figure 10.1. In this phase, two fundamental therapies are used.

10.4.2.1 Biological Treatment

Anaerobic digestion is the biological treatment used in this phase of treatment. Following digestion, the sludge is moved to the digester, a large enclosed tank where the main sludge is digested using facultative bacteria such as Proteobacteria, Firmicutes, Bacteroides, and Actinobacteria (Marín et al., 2015). Anaerobic bacteria use the organic nutrients in the sludge to generate carbon dioxide and methane gas. According to Oakley and Von Sperling (2017), the action of these microbes makes it safe to dispose of this sludge to the environment by lowering the cell biomass present and enhancing its dewaterability (eliminating water existing in it or removing moisture) for later treatment. To do this, the digester is heated to 35–37 degrees Celsius, and the digestion takes 15–28 days with optimum mixing to promote mass transfer. The digested gas contains 35% carbon dioxide and 65% methane, and it is utilized to power the treatment facility (i.e. large-scale treatment plants).

10.4.2.2 Chemical Treatment

This is also known as chemically enhanced primary therapy (CEPT) or advanced primary treatment (APT) (APT). They are used to boost sludge treatment performance (anaerobic digestion) as described by Metcalf and Eddy/AECOM (2014). The CEPT treatment used flocculation as well as traditional primary sedimentation. Flocculation is the process by which particles in water collide and attach, resulting in the formation of flakes, whereas APT treatment employs coagulation-flocculation with a high-rate lamellar settler for substantially shorter hydraulic retention durations (Metcalf and Eddy/AECOM, 2014). Before the effluent is disposed of, chemical treatment such as pH correction or pre-reduction of a high organic load may be necessary.

10.4.2.3 Secondary Phase of Treatment of Emergent Pollutant

The secondary treatment employed in the treatment of emergent pollutants in wastewater is the removal of biodegradable organic matter from wastewater. Secondary treatment is said to eliminate roughly 85% of the water's biochemical oxygen demand (BOD). The majority of their treatment is done through biological approaches. This therapy employs two types of biological systems.

10.4.2.4 Activated Sludge Treatment

This is a biological wastewater treatment that involves the mixture of microorganisms in a bioreactor. It is an aerobic method that uses a high biomass of microorganisms, mostly bacteria, protozoa, and fungus, to remove the clumped organic

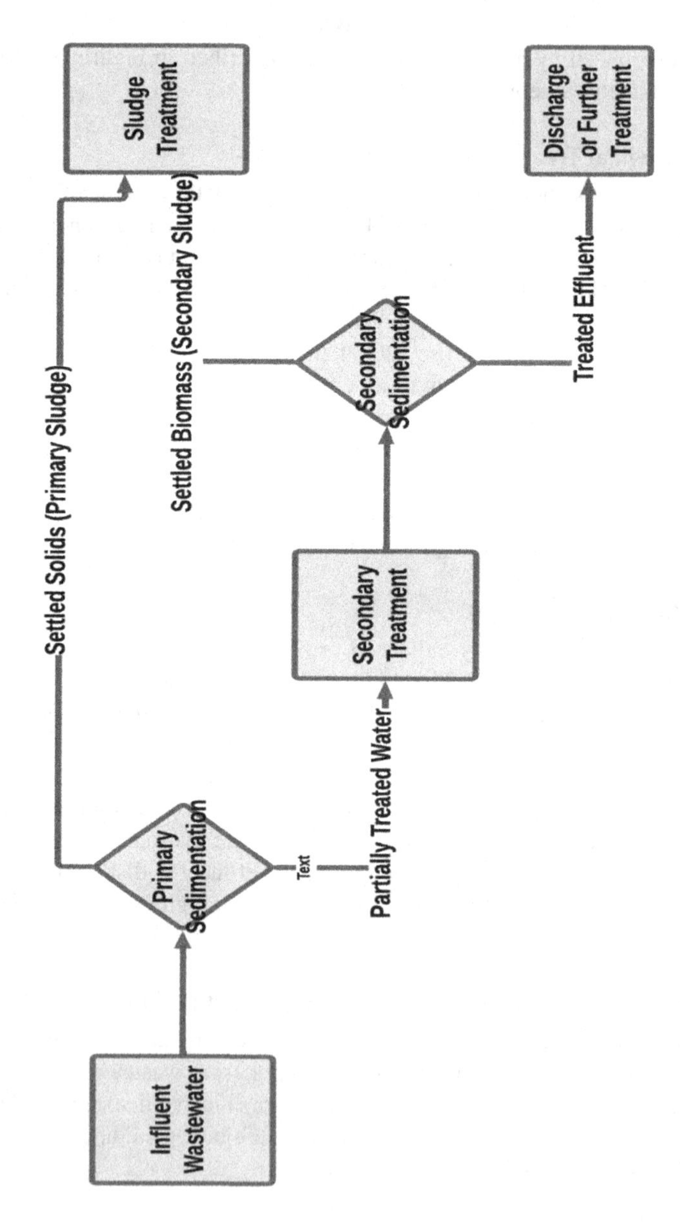

FIGURE 10.1 Description of primary sedimentation during primary and secondary treatment of wastewater and sludge management.

materials in the wastewater (Marin et al., 2015). By agitation and aeration, the main effluent from the primary treatment phase is thoroughly mixed with microorganisms as documented by Marin et al. (2015). This technology includes injecting air into the bioreactor with adequate mixing to allow oxidation to occur. The biological flocs (sedimented sludge) are then allowed to settle in a settling tank, separating the sludge from the clear treated water. A portion of the waste sludge is recycled and transferred to the aeration tank, while the remainder is collected for further treatment and disposal, either directly into the soil or by burial. Diffused aeration, surface aerators (cones), and in rare cases, pure oxygen aeration are all ways of aeration. The benefits of utilizing activation sludge treatment include a decreased ammonia level, the possibility of utilizing it in tiny places, and the fact that it is typically odor-free (Drewnowski et al., 2019). Aside from the benefits of this technology, the high energy required to operate the aeration tank has an impact on utilization (Drewnowski et al., 2019). The availability of oxygen, temperature, and the characteristics of the treated wastewater are all factors that influence the activation sludge system (Kolev, 2017).

10.4.2.5 The Use of Trickling Filters

Trickling filters treat wastewater by carrying a large population of microorganisms on a stable surface and removing organic materials aerobically with the organisms attached to a media (usually rocks or plastic). As the wastewater travels over the medium, microorganisms in the water cling to the rock, slag, or plastic surface and form a film (Naz et al., 2015a). The organic material is then digested by aerobic microbes in the slime layer's outermost region. As the slime layers thicken due to microbial activity, oxygen cannot permeate, and anaerobic organisms form and lose their capacity to adhere to the medium, causing some of the slime to slip off the filter into the wastewater (Naz et al., 2015a). This is referred to as sloughing. The biomass consumes oxygen-demanding compounds as the wastewater flows through the media, and the water leaving the media is significantly cleaner. However, some biomass sloughs off the medium and must settle in a secondary treatment tank. The subsurface drainage system collects the slime and transports it to the clarifier for disposal (Oakley and Von Sperling, 2017). The secondary treatment effluents are subjected to further treatment to minimize the microbial burden. The advantages of this system include the fact that it is appropriate for regions where large areas of land are not accessible, that it is successful in treating high concentrations of organic waste depending on the medium used, that it is durable, and that it has minimal power needs. The disadvantages of this treatment technique include the following: it necessitates the operator's continual attention, the incidence of clogging is quite high, there are vector and odor issues, and flexibility and control are restricted when compared to activated-sludge procedures (Oakley and Von Sperling, 2017)

10.4.2.6 Tertiary Phase of Treatment of Emergent Pollutants

After secondary treatment, tertiary treatment of wastewater entails several additional procedures to further eliminate pathogenic bacteria, non-biodegradable chemicals, heavy metals, and some residual inorganic dissolved solids (Hu et al., 2017). In this treatment, wastewater satisfies a higher standard than the secondary treatment's effluent. The majority of methods include some form of physicochemical treatment, such as coagulation, filtration, organic activated carbon adsorption, reverse osmosis, and further disinfection. Others involve the application of UV radiation. This is the most sophisticated therapy used to tackle emerging contaminants (Hu et al., 2017). UV radiation is one of the most regularly utilized treatment methods (Hu et al., 2017).

10.4.2.7 Ultrafiltration

Ultrafiltration is a method intermediate between microfiltration and nanofiltration. Ultrafiltration membranes typically have holes ranging from 10A to 1,000A. Larger particles move through the membrane less easily when the pore diameters are smaller. As a result, the rate of fluid filtration is determined by the pore size of the membrane (Kim et al., 2018). According to Kim et al. (2018), pressure is the driving factor for mass transfer in ultrafiltration, as it is in reverse osmosis, nanofiltration, and microfiltration. The pressure differential for the ultrafiltration process is typically in the range of 2 to 5 bars, and it is applied to the system via the pump and the feed solution. Solvent molecules and tiny dissolved solids move across the membrane under these circumstances, whereas bigger molecules and colloidal particles are trapped and remain behind them. The purpose of using this approach is to deal with the phenomena of concentration polarization and membrane fouling.

10.4.2.7.1 Nanofiltration and Microfiltration

The membranes used in nanofiltration contain pores close to or lower than a nanometer, and the size is usually around 0.5 nm to 1.5 nm. It is a process between ultrafiltration and reverse osmosis. The holes of nanofiltration membranes are near to or less than a nanometer in size, typically ranging from 0.5 nm to 1.5 nm. Kim et al. (2018) described microfiltration, on the other hand, as a filtration technique that uses porous membranes to separate suspended particles with diameters ranging from 0.9 microns to 90 microns from the flow of gases and fluids. This method's acceptable benefits include the appropriate specification of the external product (output), simple operating circumstances, the membrane process's capacity to combine with other processes (adsorption, chemical, and biological features), and the system's simpler control.

10.4.2.7.2 Reverse Osmosis

Reverse osmosis is a method that separates minerals from wastewater using membranes that are permeable to water but not to minerals (Collivignarelli et al., 2018; Kim et al., 2018). By exerting pressure from the feeding side, mineral-containing water is forced through the membrane, and solute-free water is collected as the leaking flow. Under pressure, water is forced through a succession of semipermeable membranes. External pressure exceeds normal osmotic pressure. The bigger molecules are then cleansed as they pass in a lateral flow along the membrane. Because these holes hinder the passage of molecules bigger than water molecules, as a result, the smaller molecules can pass through the pores of the membrane, whereas the bigger ones cannot (Kim et al., 2018). The bigger molecules are then cleansed as they pass in a lateral flow along the membrane. Because these pores prohibit molecules bigger than water molecules from passing through, permeated water flows from one side and concentrated water (brine) flows from the other.

Reverse osmosis can remove 90% to 99% of all TDS and colloidal particles from water. Sodium, potassium, sulfate, bicarbonate, silicate salts, bacteria, viruses, and other wastewater-soluble entities are examples of these salts. According to Kim et al. (2018), emerging contaminants from pharmaceutical goods such as ciprofloxacin may be eliminated using a reverse osmosis membrane under continuous pressure.

10.4.2.7.3 Ultraviolet Radiation

According to Collivignarelli et al. (2018), this form of treatment is employed particularly for disinfection since it does not produce undesired disinfection by-products. UV light is employed because it destroys the DNA of bacteria and also produces a photooxidation by-product that kills or damages any diseases or germs present. Photooxidation is a chemical process that happens when light is absorbed in the presence of oxygen. Some bacteria are known to be resistant to UV radiation, while others can mend and recover from its effects (Collivignarelli et al., 2018).

In terms of effluent quality, simplicity of operation and maintenance procedures, cheap cost, and the small amount of operating space required, nanofiltration is superior to other traditional filtration technologies for processing medicines and personal care items (Hu et al., 2017; Mehran Abtahi et al., 2018). However, reverse osmosis is the most effective method of pesticide removal. In actuality, there are no universal principles since performance is dependent not only on the membranes employed and the pretreatments that the effluents go through but also on several criteria such as the complexity of the effluents, the type(s) of substance(s), or their concentration. As a result, several studies must be carried out before deciding on the membrane technology to be employed.

According to Mehran et al. (2018), the upsides of membrane filtration embody the availability of a wide variety of commercial membranes from numerous manufacturers, standard style, little footprint, easy and economical technologies, production of high-quality treated waste matter, no chemicals needed, low energy consumption, and well-known separation mechanisms.

10.5 CONTEMPORARY STRATEGIES FOR WASTE REMOVAL/ MANAGEMENT FROM TREATMENT FACILITIES

10.5.1 PHYSICAL WASTE REMOVAL STRATEGIES

Physical waste removal methods employ mechanical properties for the separation and removal of solid particulate matter from wastewater.

10.5.2 SEDIMENTATION

Sedimentation is the oldest and most commonly used technique in the treatment of wastewater Zabava et al., 2016). It refers to the process by which coagulated flocs are allowed to settle. A variety of chemical compounds are utilized in facilitating this process. These compounds may include salts of metals such as ferrous sulfate ($FeSO_4.7H_2O$), ferric chloride ($FeCl_3$), ferric sulfate ($Fe_2(SO_4)_3$), aluminum sulfate

$(Al2SO_4)_3$, and calcium hydroxide ($CaOH_2$) (Bezirgiannidis et al., 2019). Numerous studies have been conducted to assess the application of sedimentation in wastewater treatment.

Bezirgiannidis et al. (2019) evaluated the use of chemically enhanced primary sedimentation/treatment (CEPT) in the treatment of municipal wastewater. After CEPT, significant improvements in pollutant parameters, including chemical oxygen demand (COD), 5-day biological oxygen demand (BOD5), nitrogen (NH4+-N), total suspended solids (TSS), volatile suspended solids (VSS), and phosphorus (PO43-P), were reported, characterized by removal efficiencies of 62%, 56%, 14%, 72%, 71%, and 71%, respectively.

In another study by Ren et al. (2020), chemical sedimentation, aimed at turbidity, organic and inorganic phosphorus removal from turbid wastewater was proposed. It involved the use of ferric chloride ($FeCl_3$) as a precipitant followed by the application of starch-based flocculant. This modified process led to a decrease in turbidity and total phosphorus of 95% and 90%, respectively, values higher than observed in traditional chemical sedimentation using just $FeCl_3$. Lin et al. (2017) developed a new strategy involving the use of $FeCl_3$ as a coagulant for chemically enhanced primary sedimentation (CEPS). It was aimed at improving the removal of nutrients, energy saving, and resource recovery in municipal wastewater treatment. At a concentration of 20 mg-Fe/L, the removal efficiencies of total organic carbon (TOC) and phosphorus (PO43-P) were 75.6% and 99.3%, respectively. Chen et al. (2019) demonstrated a dose-dependent action of $FeCl_3$ in the removal of pollutants from raw wastewater by sedimentation. 25 mg/L $FeCl_3$ was identified as the optimal dosage for pollutant recovery, as it led to about 78% of COD and 95% of phosphorus.

10.5.2.1 Filtration

Filtration is an important stage in water treatment. It involves the removal of suspended solid substances from water using various processes such as fine mesh screening, media filtration, and membrane filtration. Media filtration systems employ layers of filter media such as sand, anthracite, and gravel to separate suspended solids from incoming feed water (Liu et al., 2018). Different types of materials have been studied to determine their efficacy in the removal of various forms of waste materials from water. Liu et al. (2018) applied deep filtration processes in the removal of oils from oily wastewater using ZnO and octadecyltrichlorosilane-coated quartz sand filter media. It was observed that the filter media used in the process enabled a highly efficient filtration process and could potentially be applied in oily waste treatment. In another study, Shen et al. (2021) evaluated the efficiency of aluminosilicate filter media in the removal of microplastics from wastewater. The aluminosilicate filter media used was modified by cationic surfactant, and it was observed to have a significantly higher removal (>96%) of microplastics than rapid sand filters (63%).

Another media filtration method used in waste management is slow sand filtration (SSF), which is widely regarded as one of the most proficient strategies for the elimination of pathogens, particulate organic matter, and turbidity from wastewater (Amy et al., 2006). Kaetzl et al. (2020) evaluated the applicability of anaerobic biochar filters and slow sand filters in wastewater treatment. Their study revealed a significant extent of COD and TOC removal by both filters used but a higher removal in biochar filters than in sand filters. Additionally, biochar filters removed a significantly higher concentration of *E. coli* (1.35 ± 0.27 log10MPN·100 mL^{-1}) than sand filters (1.18 ± 0.31 log10MPN·100 mL^{-1}). In addition, Joel et al. (2018) demonstrated the use of slow sand filtration techniques in raw wastewater treatment. The filtered wastewater was observed to meet the recommended COD and TC standards set for irrigation purposes. The filtration mechanism was explained to be a combination of mechanical (absorption, diffusion, screening, and sedimentation) and biological processes as water percolated through the sand.

10.5.2.2 Solidification and Stabilization

Solidification and stabilization (S/S) are major treatment strategies for the management of industrial-generated hazardous wastes. It involves mixing a binding reagent into the contaminated media or waste (Hashemi et al., 2019). Solidification and stabilization are progressively efficient pretreatment techniques that encapsulate and immobilize hazardous waste materials to achieve a decrease in leaching of the toxic elements in landfills (Hashemi et al., 2019). It involves the application of chemical agents, in addition to water and other sludge components, to form stable solids. According to numerous publications, the most common S/S material is cement. Tajudin et al. (2016) conducted a review of various agents used in S/S processes, including ordinary Portland cement (OPC), quicklime, fly ash, rice husk ash, bottom ash, calcium aluminate cement (CAC), pozzolanic cement, sulfate-resistant Portland cement (SRC), montmorillonite, diatomite, and sepiolite, among others. In an earlier study, Malviya and Chaudhary (2006) extensively reviewed the application of solidification and stabilization processes in treating hazardous wastes. Parameters to evaluate solidification were highlighted as strength, setting time, and extent of hydration. Cement content, curing time, and water-to-solid ratio were also identified as factors known to affect the strength development of the solidified/stabilized matrix. It was further emphasized that the performance of S/S products could be enhanced using additives such as silica fume, sulfur polymer, $CaCl_2$, $Ca(OH)_2$, Na_2SO_4, and K_2SO_4.

Solidification/stabilization is described as one of the most desirable techniques for the remediation of soils polluted with heavy metals, owing to its convenience and effectiveness. Jiang et al. (2019) analyzed the application and mechanism of action of alkaline industrial wastes (soda residue, steel slag, carbide slag, and red mud) for the solidification/stabilization of soil contaminated with heavy metal. Soda residues, steel slag, and carbide slag were identified as potential agents for the solidification and stabilization of Pb, Cd, Zn, and Cu, while red mud was ascertained as a prospective agent for the stabilization of As. Hydration, precipitation, and adsorption were also noted as the mechanisms of action of the alkaline wastes. Also, a limitation in the implementation of remediation processes using these agents was underscored due to a lack of sufficient available information on the long-term effectiveness, synergistic effects, and implementation in multiple heavy metal-polluted soils.

Han et al. (2020) assessed the prospects of bacterial-induced mineralization (BIM) as an S/S strategy. The basis of its application is the production of acid anions by some bacteria which react with (heavy) metal ions in the soil to form stable metal salt precipitates. The precipitates are observed to promote consistency between solid particles in the medium, a process that is analogous to the binding effect in cement-facilitated S/S. Compared to the regular cement-based S/S procedure, BIM attained higher strength improvement capacity, water permeability, and durability. Additionally, Singh et al. (2023) evaluated the use of fly ash-based geopolymers (combinations of sodium silicate and alkaline compounds such as NaOH or KOH) as an alternative to cement for the environmentally friendly and cost-effective management strategy of hazardous wastes. The geopolymers were observed to confer significant compressive strength of 2 MPa–60 MPa in addition to total immobilization of heavy metals (such as Cr, Cd, and Pb).

10.5.3 Flotation

Flotation is a technique used in treating industrial wastes, involving the introduction of gas as the transport medium. It is applied in treating wastewater containing fine suspended solids and oily matter and is often used as an alternative to sedimentation. Types of flotation processes include dissolved air flotation, dispersed air flotation (or induced air flotation), plain gravity flotation, vacuum flotation, electrolytic flotation (or electroflotation), and biological flotation (Wang, 2006).

10.5.3.1 Dissolved Air Flotation (DAF)

Dissolved air filtration is a wastewater treatment process that involves the use of air to remove suspended matter, metals, oil, and grease. According to Rocha e Silva et al. (2018), it is a process that utilizes microbubbles injected from a saturator for the separation of suspended solids or oil droplets in water. Traditionally, DAF involves four stages: generation of microbubbles, chemical pretreatment of the wastewater, flotation, and sludge removal. The efficacy of DAF has prompted the incorporation of hybrid techniques which have had numerous applications in the removal of pollutants and adherence to environmental standards of turbidity, color, total suspended solids (TSS), chemical oxygen demand (COD), and biochemical oxygen demand (BOD) (Muñoz-Alegría et al., 2021).

One of the major applications of DAF is in the treatment of oily wastes, with oil contents that may range between 1% and 50% (Rocha e Silva et al., 2018). Studies have shown that regardless of the oil content, DAF is the best strategy for the treatment of oily wastes.

Lee et al. (2020) developed and reported a hybrid system for the treatment of industrial wastewater containing oils, volatile organic compounds (VOCs), and heavy metals. It applied principles of dissolved air flotation, electrochemical advanced oxidation process, and magnetic biochar. As a subpart of their study, a pilot-scale DAF system was employed in the treatment of laundry and food wastewater. From the results, it was observed that about 99.3% of the CODs and DOCs were removed from laundry, while 75% and 79% of the CODs and TOCs, respectively, were removed from the food wastewater.

Ortiz-Oliveros and Flores-Espinosa (2018) studied the simultaneous elimination of oil, total cobalt, and 60Co from radioactive liquid waste using dissolved air flotation. The results revealed a higher efficiency of DAF (94.84%) in the removal of oily wastes than total cobalt and 60Co (85.39% and 87.5%, respectively). Pereira et al.

(2020) evaluated the performance of flocculants Tanfloc and polyacrylamide (PAM) in the treatment of dairy wastewater using dissolved air filtration. The results showed a turbidity removal of about 90%.

Ortiz-Oliveros and Flores-Espinosa (2020) proposed the use of a mobile dissolved air flotation system (MDF) for the treatment of liquid radioactive waste. Their results revealed a 95% turbidity removal using the proposed system, along with significant removals of other parameters (SO_{4-}^{2}, NO_3–, PO_{3-}^{3}, Ni, Cr, and Pb concentrations). Total solids, Co, and 60Co removal were reported to be about 85%–88%, 94%, and 75%, respectively. Additionally, Leite et al. (2019) demonstrated the use of DAF (preceded by coagulation) in retrieving microalgae cultivated in wastewater. Significant removal of turbidity (93.7%–96.2%), apparent color (91.7%–92.3%), total Kjeldahl nitrogen (TKN) (90.2%–92.6%), total phosphorus (89.4%–90.9%), and TSS (88.6%–92.5%) were observed. Additionally, Pooja et al. (2021) researched the treatment of wastewater from the electroplating industry using dissolved air flotation. It was aimed at the removal of toxic metals such as chromium, cadmium, nickel, lead, and copper from wastewater. Optimum conditions led to a total removal of 97.39% of all metals.

10.5.3.1.1 Dispersed Air Flotation

Dispersed air flotation (or induced air flotation) is a method involving the direct introduction of air into the wastewater through a revolving impeller, a diffuser, or an ejector at a pressure slightly higher than 1 atm for the generation of air bubbles (usually between 80 microns to over 1 mm in size) in high volumes under turbulent conditions (Wang, 2006). According to Alhattab and Brooks (2017), dispersed air flotation involves the production of bubbles by the injection of compressed air through the pores of a diffuser, with a diameter generally ranging between 60 µm and 655 µm and a bubble number of 2×105/cm3. The effectiveness of the dispersed air flotation process is reported to be dependent on the level of instability of the suspended particles, particle size, and extent of particle capture by bubbles. The addition of surfactants may also influence the efficiency of the procedure (see Figure 10.2 for pictorial illustration).

10.5.3.2 Coagulation/Flocculation

Water industries consider coagulation and flocculation to be the key treatment strategies for enhancing the overall efficiency and cost-effectiveness of wastewater treatment (Cui et al., 2020). Coagulation/flocculation involves the addition of compounds (coagulants) that stimulate the aggregation of minute flocs into larger flocs so that they can be more easily separated from the liquid part. As a result of an increase in the severity of environmental pollution problems and stringency in water quality demands, conventional coagulation/flocculation techniques have fallen short in their ability to meet requirements for water safety standards. This has led to the advent of enhanced coagulation and optimized coagulation which involve improved treatment processes to achieve the effect of enhanced wastewater treatment (Cui et al., 2020).

There is already existing information on enhanced coagulation and optimized coagulation techniques provided by many authors. Some of these technologies are discussed next.

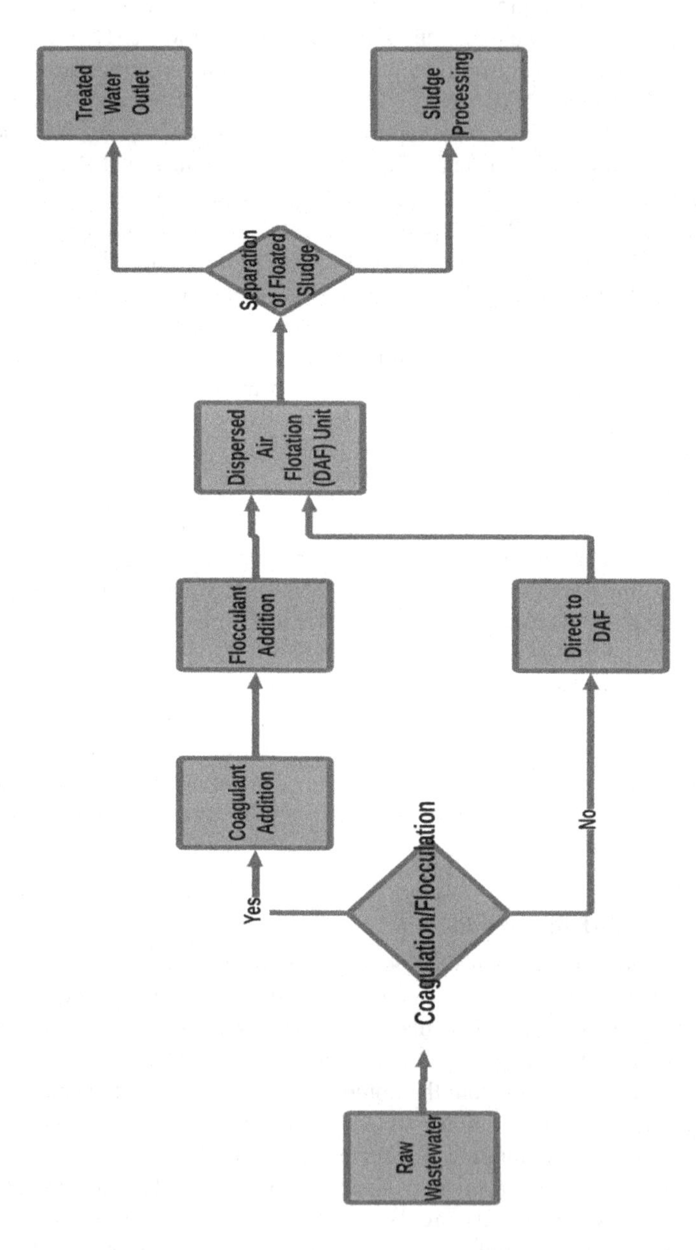

FIGURE 10.2 Flowchart describing the dispersed air flotation system in wastewater treatment.

10.5.3.2.1 Ballasted Coagulation

Ballasted coagulation as an enhanced coagulation method has been extensively applied in wastewater treatment to facilitate the removal of organic matter, bacteria, color, and algae (Zhang et al., 2021). Gaikwad and Munavalli (2019) studied the efficacy of *Strychnos potatorum* and *Moringa oleifera* with and without a ballasting agent (micro sand) in stimulating flocculation. It was observed that sand-ballasted coagulation with the coagulants (SP and MO) increases flocculation and turbidity removal by 5%–10% when compared to conventional coagulation. This finding was in line with an earlier study by Mackie et al. (2016), who reported a significant enhancement in the removal efficiency of particulate and dissolved contaminants by the action of micros and as a ballasting agent.

10.5.3.2.2 Magnetic Separation Technology

Li et al. (2022) proposed an advanced strategy for the disposal of high arsenic wastewater using nano-zerovalent iron (nZVI) and ferric oxyhydroxide (FeOOH) combined with superconducting high-gradient magnetic separation (HGMS) technology. The results showed that 99.06% of arsenic was successfully removed from wastewater through magnetic flocculation. Additionally, Prot et al. (2019) used magnetic separation techniques to separate phosphorus in the form of vivianite ($Fe(II)_3(PO4)_2.8H_2O$) from wastewater. Vivianite is reported to be the major phosphorus sink in digested sewage sludge and may contain between 70%–90% of the total phosphorus. From the results, magnetic separation, coupled with alkaline treatment, led to a recovery of 16%–32% of phosphorus from the digested sludge. It was concluded that this technology could be feasibly implemented in the extraction of phosphorus from digested sludge.

10.5.3.2.3 Chemical Oxidation-Enhanced Coagulation

Zhao et al. (2013) assessed the action of chemical oxidants ozone and potassium permanganate on coagulation in micropolluted raw landscape lake water. Compared with coagulation by polyaluminium chloride (PAC), results showed that pre-oxidation using 2 mg/L of O3 for 20 minutes could significantly enhance the elimination rate of turbidity and total organic carbon (TOC 5.33% and 5.06%, respectively. Furthermore, Dong et al. (2022) reported enhancements in the use of a combination of permanganate (Mn(VII)) pre-oxidation and Fe(III)/peroxymonosulfate (PMS) coagulation process (Mn(VII)-Fe(III)/PMS) for the removal of dissolved organic carbon (DOC), turbidity, and micropollutants from raw water. Again, Qiu et al. (2021) studied a strategy for the effective removal of *Microcystis aeruginosa* and controlling algal organic matter from eutrophic water using Fe(II)/peroxymonosulfate (PMS) pre-oxidation–enhanced coagulation. The highest removal efficacy of *M. aeruginosa* (94.3% of an initial 1.0 × 10 6 cells/mL concentration) was observed at low PMS dosages of 20 mg/L and Fe(II) dosage of 4.5 mg/L.

Other enhanced coagulation methods include ultrasound-enhanced coagulation (Huang et al., 2021), inorganic polymer flocculants (Zhang et al., 2021; Yang et al., 2018), and biopolymer-based flocculants (Makhtar et al., 2020; Xia et al., 2018; Cunha et al., 2020).

10.5.4 Biological Waste Removal Strategies

The biological waste management process employs microorganisms for the breakdown of waste materials into stable end products (Bressani-Ribeiro et al., 2018). Based on the presence of dissolved oxygen, biological waste treatment is divided into aerobic and anaerobic treatment practices.

10.5.4.1 Trickling Filter

The trickling filter (TF) is an aerobic treatment system that utilizes microorganisms attached to a medium to remove organic matter from wastewater. TFs are non-submerged aerobic biofilm reactors, consisting of a tank containing highly permeable material, onto which wastewater is applied through a distribution system (Bressani-Ribeiro et al., 2018).

Naz et al. (2015b) assessed the performance of a locally designed pilot-scale stone-media trickling biofilter for the treatment of municipal wastewater. The system was evaluated for its removal capacity of pollution indicators COD, BOD_5, NH_4, and fecal coliforms under conditions of increasing environmental temperature ranging from 20°C to 40.5°C over 40 days. The results showed a removal efficiency of 62.4%, 56.4%, and 33.8% for COD, BOD5, and NH4-N, respectively at 20°C on day one. At day 40, the removal efficiencies had increased to 98.1%, 98.6%, and 93.5% for COD, BOD5, and NH4-N, respectively, at 40.5°C. Additionally, an 88% reduction of fecal coliform was reported during the process.

Trickling filters have also been applied in the treatment of toxic raw chemicals. A study by Wu et al. (2016) was aimed at treating waste containing methyl acrylate gas. For the process, a three-layer laboratory scale biotrickling filter, packed with ceramic particles and immobilized with activated sludge, was established. The setup was designed to determine a relationship between the efficiency of the process, inlet concentration, temperature, time, and flow rate. The highest removal efficiency (>97%) was observed between day 29 and day 52 at a concentration of 4292.1 mg/m3. An increase in the flow rate at the same concentration did not lead to a decrease in the removal efficiency of the process until the 72nd day.

In a similar study, Yang et al. (2018) assessed the removal of n-butyl methacrylate from gas streams in two biotrickling filters (BTFs). One setup contained only ceramsite (BTFa), while the other contained ceramsite coated with activated carbon fiber (BTFb). The efficiencies of the two BTFs were assessed to determine the effect of the activated carbon fiber. The results revealed that the BTFb was 15.3% more efficient in the removal of n-butyl methacrylate than BTFa under empty bed residence time (EBRT) of 44s and inlet concentration of 500 mg/m3.

Zhao et al. (2022) investigated the application of a biotrickling filter packed with bamboo charcoal powder-based polyurethane (BC-PU) for the simultaneous treatment of n-hexane and dichloromethane (DCM). The highest removal capacity of n-hexane and DCM was 12.68 gm–3h–1 and 30.28 gm–3h–1 DCM, respectively. It was also observed that an increase in inlet loading (IL) of DCM led to a corresponding decrease in the removal efficiency of n-hexane. This inhibitory effect was explained to be a result of competition between enzymes produced by different microorganisms for substrate-active sites. It was concluded that BC-PU could be extensively applied in the treatment of hydrophobic pollutants.

10.5.4.2 Membrane Filtration

Membrane filtration (MF) is a pressure-driven separation procedure that uses a membrane for the separation of particles and macromolecules (Benjamin and Lawler, 2013). Membrane filtration is considered a more effective pretreatment technique than traditional methods because in comparison with the latter, membrane systems necessitate less space and chemicals (Jeong et al., 2017). Membrane filtration technology is comprised of five major treatment techniques: reverse osmosis (RO), ultrafiltration (UF), microfiltration (MF), nanofiltration (NF), and electrodialysis (ED) (Azimi et al., 2017). An aspect of membrane filtration that has received significant attention recently is direct membrane filtration. It employs a micro-/ultrafiltration membrane for the removal of solids and pathogens from wastewater, leading to the generation of easily recovered water for a variety of purposes (Ruigómez et al., 2020).

As an advantage, direct membrane processes do not require an added activated sludge stage, leading to decreased energy requirements. A downside of the application of direct membrane processes in wastewater treatment is the occurrence of membrane biofouling (Hube et al., 2020). This limitation is overcome by the removal of available nutrients for microorganisms in the feed stream (in reverse osmosis systems) or by continuous or intermittent air/gas sparging and backwashing (Jeong et al., 2017; Ruigómez et al., 2020). Giagnorio et al. (2018) assessed the viability of nanofiltration to be used as a purification method for water polluted with chromium. From their results, membrane nanofilters, NF270 and NF90, successfully achieved 98.8% and 76.5% chromium removal.

Sandoval-Olvera et al. (2019) employed ultrafiltration membrane technology for the elimination of chromium from synthetic solutions. As a result of negative charges deposited due to modifications made on the membrane, it successfully repelled different anionic solutions of nitrate, acid chromate, chromate, and a mixture of acid/diacid phosphates. Katsou et al. (2011) examined the use of a sorbent-assisted ultrafiltration (UF) system for the removal of lead, copper, zinc, and nickel from industrial wastewater. The percentage reduction was 99.9%, 99.8%, 99.7%, and 99.6% for Pb, Cu, Zn, and Ni, respectively. Additionally, the total nitrogen, TOC, and COD removal ranged from 31% to 38%, 69% to 82%, and 72% to 86%, respectively.

10.5.4.3 Activated Sludge

The activated sludge (AS) process is currently the most widely used biological wastewater treatment process in the developed world (Scholz, 2015). It is a vital step in secondary wastewater treatment and involves two main steps: aeration and sludge settlement. The activated sludge is known to contain a large number of bacteria in it, the purpose of which is to help in breaking down organic matter and also in seeding bacteria. The main principle of activated sludge is the formation of flocs by bacteria as they grow in a liquid medium (Muralikrishna and Manickam, 2017). These flocs clump together and settle at the bottom of the tank, leaving a comparatively clear supernatant, void of suspended and organic matter. A mixture of the screened wastewater and recycled liquid (containing large amounts of microorganisms) is aerated and allowed to settle. The settled solid that results is called the activated sludge (Muralikrishna and Manickam, 2017).

Lachheb et al. (2016) carried out a study aimed at decreasing the biological oxygen demand (BOD_5), chemical oxygen demand (COD), and total suspended solids (TSS) of urban wastewater using an activated sludge treatment method for reuse in irrigation. The results revealed a 90.20%, 89.79%, and 97.67% decrease in BOD_5, COD, and TSS, respectively, and a general improvement in the quality of treated water. In another study, Shivaranjani and Thomas (2017) studied the treatment of wastewater using activated sludge. From their results, BOD_5 was reduced at an efficiency of 93.7% to 40 mg/L and turbidity was reduced at 87.6% efficiency to 80 NTU. Thus, it was concluded that the activated sludge process was a very effective method for the treatment of Institutional wastewater for BOD and turbidity reduction.

Additionally, Shokoohi et al. (2020) conducted a study aimed at assessing the effectiveness of activated sludge and natural wastewater treatment systems for the treatment of municipal wastewater. Parameters (BOD_5, COD, and TSS) were measured and used to evaluate the effectiveness of the treatment methods. From the results, the COD removal efficiency of activated sludge ranged between 86.97% and 61.6%, while BOD5 removal efficiency ranged between 85.18% and 72.01%. These effluent characteristics complied with Environmental Protection Agency standards and so were deemed fit for reuse or discharge into water bodies. Furthermore, a study by Dharaskar (2015) was conducted to evaluate the viability of the activated sludge process to be used for the treatment of synthetic wastewater (containing glucose, urea, magnesium sulfate, potassium phosphate, calcium chloride, and ferric chloride). The percentage reduction in COD was used as an indicator to determine the total efficiency of the procedure. Optimum results correlating with a maximum COD removal efficiency of 94% were obtained at a mixed liquor volatile suspended solids (MLVSS) concentration of 3,000 mg/L after one hour.

El Moussaoui et al. (2019) performed an experimental pilot study evaluating the treatment of municipal sewage wastewater by an activated sludge process. The biomass growth, system performance, flocs settlement ability, and microbial hygienic parameters were determined and assessed. The results showed a percentage reduction of TSS, COD, BOD5, TKN, and TP of 92%, 94%, 97%, 85%, and 16%, respectively. The values obtained from the treated wastewater sufficiently met the regulatory requirements for direct discharge into the environment. Also of particular note was that microbiological comparison of raw and treated wastewater showed no significant decrease in total microbial count ($p>0.05$) post-treatment.

10.6 POSSIBLE CHALLENGES AND PROSPECTS FOR FURTHER STUDY

Water is a scarce natural resource, and this elevates the need for its conservation, as the increase in demand for water due to the increasing global population as well as climate change has contributed to its insufficiency. Subsequently, the conservation of water supports aquatic lives and the livelihood of a substantial amount of the earth's population while providing sources of food and clean energy for the populace. This propels the need for efficient and sustainable waste management techniques. Much attention has been drawn to the adverse effects of EPs on the environment, especially the aquatic ecosystem. To protect the environment and promote sustainability, waste management strategies usually strive to reduce challenges associated with ecological development. Presently, wastewater treatment is a challenging task since it significantly impacts the ecosystem and poses some socioeconomic challenges which are not limited potential production of secondary wastes that will require additional treatment, thereby elevating the cost-intensiveness. These wastes require comprehensive separation and removal due to their various behaviors, uneven sizes, and different physicochemical characteristics. The high cost of removing EPs via adsorption is an existing challenge. Incorporation of natural coagulation technology to the current adsorption techniques is an efficient, economical, and ecofriendly means of treating such wastes. There is a possibility of having incomplete mineralization or low mineralization efficiency, thus leading to the possible production of toxic intermediates. Finding a universal method to completely remove all contaminants from wastewater is challenging, as many biological and physical wastewater management technologies have been researched to eliminate emerging pollutants, but they have not been able to identify the most effective approach to achieve efficiency, environmental friendliness, and affordability. To improve removal and achieve desirable water quality, the prospects for effective removal of EPs from waste treatment facilities entail the implementation of innovative technology with high treatment efficiency when tackling the issues concerning emergent pollutants in wastewater. This includes combining or hybridizing effective treatment methods to achieve desired results while considering affordability and sustainability.

REFERENCES

Afshan, A., Ali, M.N. and Bhat, F.A. (2020). Environmental phthalate exposure in relation to reproduction outcomes and health endpoints. In *Handbook of Research on Environmental and Human Health Impacts of Plastic Pollution*. IGI Global, pp. 340–369.

Alhattab, M. and Brooks, M.S.-L. (2017). Dispersed air flotation and foam fractionation for the recovery of microalgae in the production of biodiesel. *Separation Science and Technology*, 52(12), 2002–2016. http://doi.org/10.1080/01496395.2017.13089

Amy, G., Carlson, K., Collins, M.R., Drewes, J., Gruenheid, S. and Jekel, M. (2006). Integrated comparison of biofiltration in engineered versus natural systems. *Recent Progress in Slow Sand and Alternative Biofiltration Processes*, 1, 3–11.

Azimi, A., Azari, A., Rezakazemi, M. and Ansarpour, M. (2017). Removal of heavy metals from industrial wastewaters: A review. *ChemBioEng Reviews*, 4(1), 37–59.

Bai, X., Lutz, A., Carroll, R., Keteles, K., Dahlin, K., Murphy, M. and Nguyen, D. (2018). Occurrence, distribution, and seasonality of emerging contaminants in urban watersheds. *Chemosphere*, 200, 133–142.

Benjamin, M.M. and Lawler, D.F. (2013). *Water Quality Engineering: Physical/Chemical Treatment Processes*. John Wiley & Sons.

Bezirgiannidis, A., Plesia-Efstathopoulou, A., Ntougias, S. and Melidis, P. (2019). Combined chemically enhanced primary sedimentation and biofiltration process for low cost municipal wastewater treatment. *Journal of Environmental Science and Health, Part A*, 1–6. http://doi.org/10.1080/10934529.2019.163384

Bressani-Ribeiro, T., Almeida, P.G.S., Volcke, E.I.P. and Chernicharo, C.A.L. (2018). Trickling filters following anaerobic sewage treatment: State of the art and perspectives. *Environmental Science: Water Research & Technology*, 4(11), 1721–1738.

Chen, Y., Lin, H., Shen, N., Yan, W., Wang, J. and Wang, G. (2019). Phosphorus release and recovery from Fe-enhanced primary sedimentation sludge via alkaline fermentation. *Bioresource Technology*. http://doi.org/10.1016/j.biortech.2019.01.094

Collivignarelli, M.C., Abbà, A., Benigna, I., Sorlini, S. and Torretta, V. (2018). Overview of the main disinfectin processes for wastewater and drinking water treatment plants. *Sustainability*, 10, 86. https://doi.org/10.3390/su100

Cui, H., Huang, X., Yu, Z., Chen, P. and Cao, X. (2020). Application progress of enhanced coagulation in water treatment. *RSC Advances*, 10(34), 20231–20244.

Cunha, C., Silva, L., Paulo, J., Faria, M., Nogueira, N. and Cordeiro, N. (2020). Microalgal-based biopolymer for nano-and microplastic removal: A possible biosolution for wastewater treatment. *Environmental Pollution*, 263, 114385.

Dharaskar, S.A. (2015). Treatment of biological waste water using activated sludge process. *International Journal of Environmental Engineering*, 7(2), 101. http://doi.org/10.1504/ijee.2015.069812

Dong, Z.Y., Lin, Y.L., Zhang, T.Y., Hu, C.Y., Pan, Y., Pan, R., Tang, Y.-L., Xu, B. and Gao, N.Y. (2022). Enhanced coagulation and oxidation by the Mn (VII)-Fe (III)/peroxymonosulfate process: *Performance and mechanisms. Water Research*, 119200.

Drewnowski, J., Remiszewska-Skwarek, A., Duda, S. and Łagód, G. (2019). Aeration process in bioreactors as the Main energy consumer in a wastewater treatment plant. Review of solutions and methods of process optimization. *Processes*, 7: 311.

El Moussaoui, T., Belloulid, M.O., Jaouad, Y., Mandi, L. and Ouazzani, N. (2019). Municipal sewage wastewater treatment by activated sludge process: Results of a pilot scale study. *Applied Journal of Environmental Engineering Science*, 5(4), 5–4.

Fawell, J. and Ong, C.N. (2012). Emerging contaminants and the implications for drinking water. *International Journal Water Resource Development*, 28(2), 247–263.

Ferguson, E.M., Allinson, M., Allinson, G., Swearer, S.E. and Hassell, K.L. (2013). Fluctuations in natural and synthetic estrogen concentra-tions in a tidal estuary in South-Eastern Australia. *Water Research*, 47, 1604–1615.

Gaikwad, V.T. and Munavalli, G.R. (2019). Turbidity removal by conventional and ballasted coagulation with natural coagulants. *Applied Water Science*, 9(5), 1–9.

Gallé, T., Bayerle, M., Pittois, D. and Huck, V. (2020). Allocating biocide sources and flow paths to surface waters using passive samplers and flood wave chemographs. *Water Research*, 173.

Giagnorio, M., Steffenino, S., Meucci, L., Zanetti, M.C. and Tiraferri, A. (2018). Design and performance of a nanofiltration plant for the removal of chromium aimed at the production of safe potable water. *Journal of Environmental Chemical Engineering*, 6(4), 4467–4475.

Gwenzi, W., Mangori, L., Danha, C., Chaukura, N., Dunjana, N. and Sanganyado, E. (2018). Sources, behaviour, and environmental and human health risks of high-technology rare earth elements as emerging contaminants. *Science Total Environment*, 636, 299–313.

Han, L., Li, J., Xue, Q., Chen, Z., Zhou, Y. and Poon, C.S. (2020). Bacterial-induced mineralization (BIM) for soil solidification and heavy metal stabilization: A critical review. *Science of the Total Environment*, 140967. http://doi.org/10.1016/j.scitotenv.2020.14

Hashemi, S.S.G., Mahmud, H.B., Ghuan, T.C., Chin, A.B., Kuenzel, C. and Ranjbar, N. (2019). Safe disposal of coal bottom ash by solidification and stabilization techniques. *Construction and Building Materials*, 197, 705–715. http://doi.org/10.1016/j.conbuildmat.2018

Hu, Y., Peng, Y., Liu, W., Zhao, D. and Fu, J. (2017). Chapter 4—Removal of emerging contaminants from water and wastewater using nanofltration technology. In S.H. Joo (ed.),

Applying Nanotechnology for Environmental Sustainability. https://doi.org/10.4018/978-1-5225-0585-3.ch004.
Huang, Y., Ding, S., Li, L., Liao, Q., Chu, W. and Li, H. (2021). Ultrasound-enhanced coagulation for *Microcystis aeruginosa* removal and disinfection by-product control during subsequent chlorination. *Water Research*, 201, 117334. https://doi.org/10.1016/j.watres.2021.11733
Hube, S., Eskafi, M., Hrafnkelsdóttir, K.F., Bjarnadóttir, B., Bjarnadóttir, M.Á., Axelsdóttir, S. and Wu, B. (2020). Direct membrane filtration for wastewater treatment and resource recovery: A review. *Science of the Total Environment*, 710, 136375. https://doi.org/10.1016/j.scitotenv.2019.13
Jeong, S., Naidu, G., Leiknes, T. and Vigneswaran, S. (2017). 4.3 membrane biofouling: Biofouling assessment and reduction strategies in seawater reverse osmosis desalination. *Comprehensive Membrane Science and Engineering*, 48–71. https://doi.org/10.1016/b978-0-12-409547-2
Jiang, H., Qi, Z., Yilmaz, E., Han, J., Qiu, J. and Dong, C. (2019). Effectiveness of alkali-activated slag as alternative binder on workability and early age compressive strength of cemented paste backfills. *Construction and Building Materials*, 218, 689–700.
Jiang, J.Q., Zhou, Z. and Sharma, V.K. (2013). Occurrence, transportation, monitoring and treatment of emerging micropollutants in waste water- a review from global views. *Microchemical Journal*, 110, 292–300.
Joel, C., Mwamburi, L.A. and Kiprop, E.K. (2018). Use of slow sand filtration technique to improve wastewater effluent for crop irrigation. *Microbiology Research*, 9(1), 7269.
Kaetzl, K., Lübken, M., Nettmann, E., Krimmler, S. and Wichern, M. (2020). Slow sand filtration of raw wastewater using biochar as an alternative filtration media. *Scientific Reports*, 10(1), 1–11.
Kanakaraju, D., Glass, B.D. and Oelgemoller, M. (2014). Titanium dioxide photocatalysis for pharmaceutical wastewater treatment. *Environmental Chemical Research*, 12(1), 27–47.
Katsou, E., Malamis, S. and Haralambous, K.J. (2011). Industrial wastewater pre-treatment for heavy metal reduction by employing a sorbent-assisted ultrafiltration system. *Chemosphere*, 82(4), 557–564. https://doi.org/10.1016/j.chemosphere.2010
Kim, S., Chu, K.H., Al-Hamadani, Y.A.J., Park, C.M., Jang, M., Kim, D.H., Yu, M., Heo, J. and Yon, Y. (2018). Removal of contaminants of emerging concern by membranes in water and wastewater: A review. *Chemical Engineering Journal*, 335, 896–914. https://doi.org/10.1016/j.cej.2017.11.044
Kolev, S.A. (2017). Dairy wastewaters—General characteristics and treatment possibilities—A review. *Food Technology and Biotechnology*, 55, 14–28.
Kolpin, D.W., Furlong, E.T., Meyer, M.T., Thurman, E.M., Zaugg, S.D., Barber, L.B. and Buxton, H.T. (2002). Pharmaceuticals, hormones, and other organic waste water contaminants in US streams. A national reconnaissance. *Environmental Science Technology*, 36(6), 1202–1211.
Lachheb, A., Idrissi, Y.A., Zouhri, N., Belhamidi, S., Taky, M., El Amrani, M. and Elmidaoui, A. (2016). Application of activated sludge for urban wastewater treatment and reuse for irrigation in Kenitra, Morocco. *American Journal of Applied Chemistry*, 4(2), 33–39.
Lee, J., Cho, W.-C., Poo, K.-M., Choi, S., Kim, T.-N., Son, E.-B., Choi, Y.-J., Kim, Y.M. and Chae, K.-J. (2020). Refractory oil wastewater treatment by dissolved air flotation, electrochemical advanced oxidation process, and magnetic biochar integrated system. *Journal of Water Process Engineering*, 36, 101358. http://doi.org/10.1016/j.jwpe.2020.10135

Leite, L. de S., Hoffmann, M.T. and Daniel, L.A. (2019). Coagulation and dissolved air flotation as a harvesting method for microalgae cultivated in wastewater. *Journal of Water Process Engineering*, 32, 100947. http://doi.org/10.1016/j.jwpe.2019.100947

Li, Y., Li, S., Hu, B., Zhao, X. and Guo, P. (2022). FeOOH and nZVI combined with superconducting high gradient magnetic separation for the remediation of high-arsenic metallurgical wastewater. *Separation and Purification Technology*, 285, 120372.

Lin, L., Li, R., Li, Y., Xu, J. and Li, X. (2017). Recovery of organic carbon and phosphorus from wastewater by Fe-enhanced primary sedimentation and sludge fermentation. *Process Biochemistry*, 54, 135–139. http://doi.org/10.1016/j.procbio.2016.12.0

Lin, Y., Li, T. and Zhang, Y. (2024). Effects of two typical quinolone antibiotics in the marine environment on *Skeletonema costatum*. *Frontiers in Marine Science*, 11, 1335582. http://doi.org/10.3389/fmars.2024.1335582

Liu, J., Zhu, X., Zhang, H., Wu, F., Wei, B. and Chang, Q. (2018). Superhydrophobic coating on quartz sand filter media for oily wastewater filtration. *Colloids and Surfaces A: Physicochemical and Engineering Aspects*, 553, 509–514.

Mackie, A.L., Laliberte, M. and Walsh, M.E. (2016). Comparison of single and two-stage ballasted flocculation processes for enhanced removal of arsenic from mine water. *Journal of Environmental Engineering*, 142, 04015062.

Mahmood, T., Momin, S., Ali, R., Naeem, A. and Khan, A. (2022). Technologies for removal of emerging contaminants from wastewater. *Intechopen*. DOI: 10.5772/intechopen.104466.

Maletz, S., Floehr, T., Beier, S., Brouwe, A., Behnisch, P., Higley, E., Giesy, J.P., Hecker, M., Gebhardt, W. and Linnemann, V. (2013). Invitro characterization of the effectiveness of enhanced sewage treatment processes to eliminate endocrine activity of hospital effluents. *Water Research*, 47(4), 1545–1557.

Malviya, R. and Chaudhary, R. (2006). Factors affecting hazardous waste solidification/stabilization: A review. *Journal of Hazardous Materials*, 137(1), 267–276.

Marín, I., Goñi, P., Lasheras, A.M. and Ormad, M.P. (2015). Efficiency of a Spanish wastewater treatment plant for removal potentially pathogens: Characterization of bacteria and protozoa along water and sludge treatment lines. *Ecological Engineering*, 74, 28–32.

Mehran Abtahi, S., Ilyas, S., Joannis Cassan, C., Albasi, C. and Vos, W.M. (2018). Micropollutants removal from secondary-treated municipal wastewater using weak polyelectrolyte multilayer based nanofltration membranes. *Journal of Membrane Science*, 548, 654–666.

Metcalf and Eddy (2014). *Wastewater Engineering: Treatment and Resource Recovery*, 5th ed. McGraw-Hill.

Mohd Makhtar, N.S., Idris, J., Musa, M., Andou, Y., Ku Hamid, K.H. and Puasa, S.W. (2020). Plant-based tacca leontopetaloides biopolymer flocculant (TBPF) produced high removal of heavy metal ions at low dosage. *Processes*, 9(1), 37.

Mompelat, S., Le Bot, B. and Thomas, O. (2009). Occurrence and fate of pharmaceutical products and by-products, from resource to drinking water. *Environmental International Research*, 35(5), 803–814.

Muñoz-Alegría, J.A., Muñoz-España, E. and Flórez-Marulanda, J.F. (2021). Dissolved air flotation: A review from the perspective of system parameters and uses in wastewater treatment. *TecnoLógicas*, 24(52), 281–303.

Muralikrishna, I.V. and Manickam, V. (2017). Wastewater treatment technologies. *Environmental Management*, 249–293. http://doi.org/10.1016/b978-0-12-811989-1

Naz, I., Saroj, D.P., Mumtaz, S., Ali, N. and Ahmed, S. (2015a). Assessment of biological trickling filter systems with various packing materials for improved wastewater treatment. *Environmental Technology*, 36, 424–434.

Naz, I., Ullah, W., Sehar, S., Rehman, A., Khan, Z.U., Ali, N. and Ahmed, S. (2015b). Performance evaluation of stone-media pro-type pilot-scale trickling biofilter system for municipal wastewater treatment. *Desalination and Water Treatment*, 57(34), 15792–15805. http://doi.org/10.1080/19443994.2015.10811

Oakley, S. and von Sperling, M. (2017). *Media Filters: Trickling Filters and Anaerobic Filters*. Michigan State University.

Ortiz-Oliveros, H.B. and Flores-Espinosa, R.M. (2018). Simultaneous removal of oil, total Co and 60Co from radioactive liquid waste by dissolved air flotation. *International Journal of Environmental Science and Technology*, 16(7), 3679–3686. http://doi.org/10.1007/s13762-018-1984-4

Ortiz-Oliveros, H.B. and Flores-Espinosa, R.M. (2020). Design of a mobile dissolved air flotation system with high rate for the treatment of liquid radioactive waste. *Process Safety and Environmental Protection*. http://doi.org/10.1016/j.psep.2020.07.016

Parra-Saldivar, R., Castillo-Zacarías, C., Bilal, M., Iqbal, H.M.N. and Barceló, D. (2020). Sources of pharmaceuticals in water. In *The Handbook of Environmental Chemistry*. Springer, pp. 33–47.

Patel, N., Khan, M.D., Shahane, S., Rai, D., Chauhan, D., Kant, C. and Chadhary, V.K. (2020). Emerging pollutants in aquatic environment: Source, effect and challenges in biomonitoring and bioremediation-a review. *Winter*, 6(1), 99–113.

Pereira, M. do S., Borges, A.C., Muniz, G.L., Heleno, F.F. and Faroni, L.R.D. (2020). Dissolved air flotation optimization for treatment of dairy effluents with organic coagulants. *Journal of Water Process Engineering*, 36, 101270. http://doi.org/10.1016/j.jwpe.2020.101270

Pooja, G., Kumar, P.S., Prasannamedha, G., Varjani, S. and Vo, D.-V.N. (2021). Sustainable approach on removal of toxic metals from electroplating industrial wastewater using dissolved air flotation. *Journal of Environmental Management*, 295, 113147. http://doi.org/10.1016/j.jenvman.2021.11314

Pradit, S., Noppradit, P., Sornplang, K., Jitkaew, P., Kobketthawin, T., Nitirutsuwan, T. and Muenhor, D. (2024). Microplastics and heavy metals in the sediment of Songkhla Lagoon: Distribution and risk assessment. *Frontiers in Marine Science*, 10, 1292361. http://doi.org/10.3389/fmars.2023.1292361

Prot, T., Nguyen, V.H., Wilfert, P., Dugulan, A.I., Goubitz, K., De Ridder, D.J., Korving, L., Rem, P., Bouderbala, A., Witkamp, G.J. and van Loosdrecht, M.C.M. (2019). Magnetic separation and characterization of vivianite from digested sewage sludge. *Separation and Purification Technology*, 224, 564–579. https://doi.org/10.1016/j.seppur.2019.05.057

Qiu, Y., Luo, Y., Zhang, T., Du, X., Wang, Z., Liu, F. and Liang, H. (2021). Comparison between permanganate pre-oxidation and persulfate/iron (II) enhanced coagulation as pretreatment for ceramic membrane ultrafiltration of surface water contaminated with manganese and algae. *Environmental Research*, 196, 110942.

Ren, J., Li, N., Wei, H., Li, A. and Yang, H. (2020). Efficient removal of phosphorus from turbid water using chemical sedimentation by FeCl3 in conjunction with a starch-based flocculant. *Water Research*, 170, 115361. http://doi.org/10.1016/j.watres.2019.115361

Rocha e Silva, F.C.P., Rocha e Silva, N.M.P., Luna, J.M., Rufino, R.D., Santos, V.A. and Sarubbo, L.A. (2018). Dissolved air flotation combined to biosurfactants: A clean and efficient alternative to treat industrial oily water. *Reviews in Environmental Science and Bio/Technology*. http://doi.org/10.1007/s11157-018-9477-y

Ruigómez, I., González, E., Rodríguez-Gómez, L. and Vera, L. (2020). Direct membrane filtration for wastewater treatment using an intermittent rotating hollow fiber module. *Water*, 12(6), 1836. https://doi.org/10.3390/w12061836

Sandoval-Olvera, I.G., González-Muñoz, P., Palacio, L., Hernández, A., Ávila-Rodríguez, M. and Prádanos, P. (2019). Ultrafiltration membranes modified by PSS deposition and plasma treatment for Cr(VI) removal. *Separation and Purification Technology*, 210, 371–381. http://doi.org/10.1016/j.seppur.2018.08.02

Scholz, M. (2015). *Wetland systems to control urban runoff.* Elsevier.

Shen, M., Hu, T., Huang, W., Song, B., Zeng, G. and Zhang, Y. (2021). Removal of microplastics from wastewater with aluminosilicate filter media and their surfactant-modified products: Performance, mechanism and utilization. *Chemical Engineering Journal*, 421, 129918.

Shivaranjani, S.K. and Thomas, L.M. (2017). Performance study for treatment of institutional wastewater by activated sludge process. *International Journal of Civil Engineering and Technology*, 8.

Shokoohi, R., Dargahi, A. and Karami, A. (2020). A survey on efficiency of natural wastewater treatment systems and activated sludge for municipal wastewater treatment. *Journal of Environmental Science and Technology*, 22(1), 15–25. http://doi.org/10.22034/jest.2018.15742.2426

Singh, R.P., Vanapalli, K.R., Cheela, V.R.S., Peddireddy, S.R., Sharma, H.B. and Mohanty, B. (2023). Fly ash, GGBS, and silica fume based geopolymer concrete with recycled aggregates: Properties and environmental impacts. *Construction and Building Materials*, 378, 131168.

Srivastav, A.L., Patel, N. and Chaudhary, V.K. (2020). Disinfection by-products in drinking water: Occurrence, toxicity and abatement. *Environmental Pollution*, 267, 115474.

Srivastava, P., Abbassi, R., Yadav, A., Garaniya, V., Asadnia, M., Lewis, T. and Khan, S.J. (2021). Influence of applied potential on treatment performance and clogging behaviour of hybrid constructed wetland-microbial electrochemical technologies. *Chemosphere*, 284, 131296.

Stalter, D., O'Malley, E., von Gunten, U. and Escher, B.I. (2020). Mixture effects of drinking water disinfection by-products: Implications for risk assessment. *Environmental Science*, 6, 2341–2351.

Tajudin, S.A.A., Azmi, M.A.M. and Nabila, A.T.A. (2016). Stabilization/solidification remediation method for contaminated soil: A review. *IOP Conference Series: Materials Science and Engineering*, 136, 012043. http://doi.org/10.1088/1757-899x/136/1/012043

Tremblay, L.A., Gadd, J.B. and Northcott, G.L. (2018). Steroid estrogens and estrogenic activity are ubiquitous in dairy farm watersheds regardless of effluent management practices. *Agricultural Ecosystem. Environment*, 253, 48–54.

Vulliet, E. and Cren-Olive, C. (2011). Screening of pharmaceuticals and hormones at the regional scale, in surface and ground waters intended to human consumption. *Journal of Environmental Pollution*, 159(10), 2929–2934.

Wang, L.K. (2006). Adsorptive bubble separation and dispersed air flotation. *Advanced Physicochemical Treatment Processes*, 81–122. http://doi.org/10.1007/978-1-59745-029-4_3

Wiest, L., Gosset, A., Fildier, A., Libert, C., Hervé, M., Sibeud, E., Giroud, B., Vulliet, E., Bastide, T., Polomé, P. and Perrodin, Y. (2021). Occurrence and removal of emerging pollutants in urban sewage treatment plants using LC-QToF-MS suspect screening and quantification. *Science of the Total Environment*, 774, 145779.

World Health Organization (2015). *WHO Multi-Country Survey Reveals Widespread Public Misunderstanding about Antibiotic Resistance.* World Health Organization.

Wu, H., Yin, Z., Quan, Y., Fang, Y. and Yin, C. (2016). Removal of methyl acrylate by ceramic-packed biotrickling filter and their response to bacterial community. *Bioresource Technology*, 209, 237–245. http://doi.org/10.1016/j.biortech.2016.03

Xia, X., Liang, Y., Lan, S., Li, X., Xie, Y. and Yuan, W. (2018). Production and flocculating properties of a compound biopolymer flocculant from corn ethanol wastewater. *Bioresource Technology*, 247, 924–929.

Yang, Z., Liu, J., Zhang, Y., Qin, Y., Xing, Y. and Li, J. (2018). Effective removal of n-butyl methacrylate in bio-trickling filter packed with ceramsite coated with activated carbon fibers. *International Biodeterioration & Biodegradation*, 133, 221–229. http://doi.org/10.1016/j.ibiod.2018.02.005

Zabava, B.S., Ungureanu, N., Vladut, V., Dinca, M., Voicu, G. and Ionescu, M. (2016). Experimental study of the sedimentation of solid particles in wastewater. *Annals of the University of Craiova-Agriculture, Montanology, Cadastre Series*, 46(2), 611–617.

Zhang, W., Tang, M., Li, D., Yang, P., Xu, S. and Wang, D. (2021). Effects of alkalinity on interaction between EPS and hydroxy-aluminum with different speciation in wastewater sludge conditioning with aluminum based inorganic polymer flocculant. *Journal of Environmental Sciences*, 100, 257–268. http://doi.org/10.1016/j.jes.2020.05.016

Zhao, H., Wang, L., Zhang, Z.H., Zhao, B., Zhang, H.W., Zhang, H. and Ma, R. (2013). Study on chemical pre-oxidation enhanced coagulation for micro-polluted raw water treatment. *Advanced Materials Research*, 777, 472–475.

Zhao, M., Hu, L., Dai, L., Wang, Z., He, J., Wang, Z., He, J., Wang, Z., Chen, J., Hrynsphan, D. and Tatsiana, S. (2022). Bamboo charcoal powder-based polyurethane as packing material in biotrickling filter for simultaneous removal of n-hexane and dichloromethane. *Bioresource Technology*, 345, 126427.

Zhu, H. and Kannan, K. (2019). Distribution and partitioning of perfluoroalkyl carboxylic acids in surface soil, plants, and earthworms at a contaminated site. *Science Total Environment*, 647, 954–961.

11 Economic Management and Challenges Faced in the Management of Emergent Pollutant

Solomon Ugochukwu Okom, Ebere Mary Eze, Peter Mudiaga Etaware, Precious Onome Obiebi, Uduenevwo Francis Evuen, Joel Okpoghono, Paul Ikechuku Ogwezzy, Irene Ebosereme Ainyanbhor, Oke Aruoren, Osikemekha Anthony Anani, and Joshua Othuke Orogu

11.1 INTRODUCTION

Emerging pollutants (EPs) have been identified as toxic substances in recent research conducted across the world and as perceived threat(s) to the environment and human health, with inadequate published health criteria (Pokkiladathu et al., 2022). The search for these toxic substances has become a global concern based on the high level of danger(s) posed to the environment and their lethal nature to human health. These pollutants include a variety of compounds, such as antibiotics, drugs, steroids, endocrine disruptors, hormones, industrial additives, chemicals, microbeads, and microplastics (Peña-Guzmán et al., 2019). These EPs can persist in the environment over time since they are capable of bioaccumulating in the environment with the tendency to cause potential life-threatening ailments or diseases, such as abnormal growth, reduced fertility and reproductive health issues, neurodevelopmental delays, constraining wildlife species diversity, degrading aquatic ecosystems, and possibly harming the human immune system (Han et al., 2018). The majority of emerging pollutants are not new or recent toxic substances introduced into the environment. Most emerging pollutants can cause diseases that are novel with undocumented harmful effects and unestablished modes of action(s). The term "emerging pollutants" applies to both the contaminants and the issues or problems associated with their existence in the environment.

So the present study gave an insight into the economic management and challenges faced in the management of emerging pollutants in our environment, which recently received much attention worldwide. In this article, peer-reviewed scientific

DOI: 10.1201/9781003362975-11

literature on emerging pollutants was reviewed with special attention to sources and classes of emergent chemicals, global state-of-the-art evaluation, challenges posed by emergent pollutants, and possible mitigation strategies employed in the economic management of these EPs. Therefore, this study was set up to underscore the global economic management of EPs to address the challenges faced by different countries in the management of EPs and other perceived pollutants that are lethal to the environment and deadly to all forms of life, most especially that of humankind.

11.2 ECONOMIC MANAGEMENT OF CHEMICALS, ESPECIALLY THE EMERGENT ONES

Emergent chemicals have been known for decades to elicit serious harm to biological organisms and alter the natural state of the environment. To this end, Kassotis et al. (2020) examined the implications of EDCs (endocrine-disrupting chemicals) as well as the policies and regulatory and economic impacts on the US population. In their study, pesticides, one of the ECs, can induce cancer in humans when exposed at a minimum concentration.

In the USA, screening and testing programs are focused on estrogenic EDCs exclusively, and regulation is strictly risk-based. Minimization of human exposure is unlikely without a clear overarching definition for EDCs and relevant pre-marketing test requirements. We call for a multifaceted international program (e.g. modeled on the International Agency for Research in Cancer) to address the effects of EDCs on human health—an approach that would proactively identify hazards for subsequent regulation (Kassotis et al., 2020).

In the past two years, there has been an alarming increase in the introduction of emerging pollutants into our surroundings. As long as the temperature and pressure are standard, ECs are common in the environment because they dissolve quickly without evaporating at standard levels. ECs can have detrimental effects on aquatic life and humans, including altered mineral composition, altered metabolic pathways, and altered reproductive organs, even at low concentrations (Varsha et al., 2022).

According to Zahmatkesh et al. (2022), natural specimens with continuously improved assay methods are better suited for the detection of emerging pollutants. Ozonation and chlorination seem to be the best disinfection techniques in terms of eliminating pathogens as well as newly emerging contaminants. Countries have been using ozonation and adsorption on activated carbon for approximately ten years due to its economic benefits, simplicity, and efficiency on an industrial scale. Not only can these technologies be effortlessly incorporated into current treatment facilities but they are also reasonably priced. Several biological techniques, including artificial wetlands, biomembrane reactors, biotechnology, and enzymatic degradation, are now available as a result of ongoing research efforts.

Over the past ten years, China has significantly improved the quality of its air and water thanks to coordinated efforts by the government, businesses, and civil society (Liu et al., 2023). There has been a significant decrease in the portion of pollution that emits sensory cues, like bad odor, reduced visibility, and black smoke. The water is becoming clearer, and the sky is becoming bluer. The pollution of new pollutants, which is the less obvious aspect of pollution, has drawn more attention in recent years. POPs, EDCs, and antibiotics are the primary new pollutants that China is worried about (Fan et al., 2021). Although most of these new pollutants are invisible when they are present, many of them are hazardous, persistent, and bioaccumulative. Their toxicities range from neurotoxicity to carcinogenicity, endocrine disruption, reproductive and developmental toxicity, and aquatic ecotoxicity, posing progressively greater risks to the environment and public health (Mishra and Mohanty, 2023).

The manufacture, use, and disposal of chemicals with significant economic value are the main causes of new pollutants, which have drawn more attention in China in recent years. A high level of policy coordination throughout the chemical life cycle and a balanced consideration of several social goals are necessary for managing new pollutants. We examined the present regulatory framework for controlling novel pollutants from the standpoint of policy integration. According to Liu and Wang (2020), some of the major findings are as follows: (1) there are instruments and guiding principles for balancing social, environmental, and economic objectives that can be used directly or after revision to guide the management of new pollutants; (2) new regulations are desperately needed to prevent, minimize, and clean up new pollutants throughout their entire production, use, and disposal process; and (3) complementary policies, such as information-based policies and policies for facilitating research and development, are rather weak.

11.3 CLASS OF EMERGENT CHEMICALS AND THEIR ENVIRONMENTAL AND HEALTH RISKS MANAGEMENT

11.3.1 Perfluorinated Compounds (PFCs)

PFCs are a particular type of material found in many products, including textiles, acrylics, food containers, fixatives, etc. Bai et al. (2018) reported that perfluorocarbon sulfonic acids (PFSAs) and perfluorocarbon cyanamides (PFCAs) are the two most prevalent types. However, the most well-known uses are perfluorooctanoic acid (PFOA) and perfluoro octane sulfonate (PFOS). PFOA has been found in processing and potable water as a result of emissions from non-point sources, WWTPs, and industrial sites. Due to the contaminated groundwater plume, it may also be found in drinking water wells. Comparably, PFOA from adjacent industrial sites may infiltrate the air, settle on the soil's surface, and then seep into the groundwater. According to Pitter et al. (2020), there has also been evidence that potable water sources approximately 20 miles away from the target site were affected when PFOAs were released into the environment by an industrial facility without first treating them.

11.3.2 Toxic Chemicals

Negligence can lead to the release of toxic chemicals from pesticides, fungicides, and insecticides into drinking water through reverse siphoning, field and golf course spraying, and large-scale unintentional spills (Srivastav et al., 2020).

Pesticide metabolites have been absorbed by the soil with ease (Nováková et al., 2020). Pesticide metabolites were detected in both surface and groundwater, and there was variation in the runoff specimens in sparsely populated urban regions in response to different pesticide treatments (Nováková et al., 2020). According to several compounds found in water matrices, pesticide usage is daily in urban areas (Gallé et al., 2020).

11.3.3 Hormones

Estrogens and androgens can influence physiological systems and functions in both humans and animals, as well as initiate important bodily administrative processes. Steroid hormones are naturally estrogen-producing substances secreted by the adrenal cortex and other bodily regions (Bayabil et al., 2022). The main source of these hormones is crop waste. Moreover, estrogens have been discovered in soil, rivers that empty into cow grazing areas, washout from the treatment of dairy waste, and groundwater under untreated outflow holding ponds (Mishra and Mohanty, 2023). According to Ng et al. (2021), livestock releases estrogen in rather large concentrations. It has long been believed that the natural estrogens released from human and animal feces represent a serious environmental risk. Because animal feces and other bio-solids are commonly used in organic farming, this ecological disaster is especially concerning.

Emergent pollutants “EPs” are produced by a variety of sources (Table 11.1), including pesticides that contaminate drinking water as a result of negligence, including back siphoning, applying pesticides in lawns and golf courses, and a sizable accidental spill. Perfluorinated compounds (PFCs) are applied in paints, food packaging, textiles, adhesives, polishes, waxes, electronics, and stain repellents. It was shown in Table 11.1 that the major sources of introduction of these EPs into the environment are through indiscriminate disposal of domestic, office, hospital, industrial, market, and agricultural wastes.

11.4 A STATE-OF-ART EVALUATION OF EMERGING POLLUTANTS GLOBALLY

To evaluate the detrimental impacts of possible toxicants on inhabitants and their surroundings, detention, and monitoring of developing pollutants are essential. Researchers are interested in discovering sensible solutions to the state of the art of evaluating new contaminants since it is a key approach to pollution cleanup, and environmental safety and security is a major concern globally.

Chemical analysis techniques like gas chromatography (GC), inductively coupled plasma-mass spectrometry (ICP-MS), and high-performance liquid chromatography (HPLC) have long been used to assess the sensitivity and accuracy of emerging pollutants in soil and freshwater sample analyses (Zhang et al., 2021). These techniques require a lot of work, many processes in sample preparation, hazardous chemicals, and time; also, the equipment requires operators with the necessary training. Researchers have created a more sophisticated biological approach (biosensors)

TABLE 11.1
Some Identified Sources of Introduction of Emergent Pollutants into the Environment.

No.	Type	Source	Entry route into the environment	References
1.	Perfluorinated compounds (PFCs)	Paints Textiles Adhesives Polishes Waxes Electrical gadgets Stain repellants Food packaging materials	Domestic wastes Market waste Office waste Hospital waste Industrial waste	Cheng et al. (2020)
2.	Toxic chemicals	Fungicides Pesticides Insecticides Herbicides Bactericides Germicides	Agricultural waste Industrial waste Runoffs from farm irrigation	Jacob et al. (2020)
3.	Steroids or sex hormones	Progesterone Testosterone Estrogen Plant-based hormones	Clinics Hospitals Farms Research institutes	Costa et al. (2020)
4.	Surfactants	Phthalates Polybrominated compounds (PBCs) Polychlorinated biphenyls (PCBs) Bisphenol A (BPA) Alkylphenol ethoxylates Alkylphenols	Pharmaceutical products Industrial waste Research institutes Laboratories Hospitals Clinics	Mpatani et al. (2020)

to assess new contaminants with low cost, low energy consumption, and real-time practicality to comprehend the state of the art of evaluating such pollutants (Zhang et al., 2021).

According to Huang et al. (2023), biosensors are biological tools that are used to assess developing contaminants based on their biotechnology. They include components for signal transducers that produce detectable or measurable signals while detecting pollutants. They have been effectively used for particular detections of heavy metals and organic contaminants. Numerous studies have effectively examined the hazardous and bioavailable components of metals and organic pollutants, such as cadmium and toluene (Hui et al., 2022). Biosensors have also been used to assess new contaminants in our environment, such as phenanthrene and mercury (He et al., 2018). Furthermore, it might be helpful to identify organic contaminants that resemble benzene by using the benzene metabolizing regulatory protein (Rasheed et al., 2019). The interplay between those regulatory systems. Therefore, the selection of regulatory systems or recognizing elements would have a significant impact on the selectivity of specific biosensors.

Gomes et al. (2021) demonstrated the use of carbon screen-printed electrodes for the simultaneous identification of estradiol, paracetamol, and hydroquinone in tap water. Their findings could have a significant impact on wastewater analysis. A luminescent sensor derived from a stable europium (III) metal-organic framework also showed promise for use in water quality analysis. It was subjected to antibiotic identification testing (Li et al., 2022).

Apart from using sophisticated analytical methods that researchers have made possible, it is possible to identify new emerging pollutants and the products of their transformation. Numerous international projects have been launched to assess new pollutants. For instance, the United Nations Environment Programme (UNEP) has created a framework for evaluating newly emerging pollutants. According to Mishra and Mohanty (2023), this framework entails identifying possible sources of emerging pollutants, gathering information on their prevalence and concentration, and assessing any potential effects they may have on the environment and public health.

11.5 CHALLENGES FACED IN THE MANAGEMENT OF EMERGING POLLUTANTS

The quality of water resources, wildlife, and vegetation is being negatively impacted by the ongoing introduction of new pollutants into the ecosystem. To learn more about these pollutants' toxicity, effects on the environment, and behaviors in various aquatic environments, numerous studies have been carried out all over the world. Persistent chemicals such as pesticides, industrial additives, and medications can contaminate water bodies. These contaminants can enter through a variety of openings, surpass permissible limits, and build up to cause harm to both human populations and the environment (Gani et al., 2021).

Humans and animals are greatly impacted by emerging pollutants; one effect of endocrine disruption is that it modifies, mimics, or suppresses the action of hormones within the endocrine system (Ahmed et al., 2021). It has been demonstrated that exposure to EDs causes reproductive problems, a decrease in the number of male sperms, an increase in testicular, prostate, ovarian, and breast cancers, and more (Ahmed et al., 2021). Furthermore, alterations in reproductive anatomy, fertility, eggshell thinning, and hormonal activity are some of the most obvious effects of EDs on animals (Faheem and Bhandari, 2021). Pesticides and pharmaceuticals can also affect how a species grows, reproduces, and evolves in its natural habitat. For example, it has been demonstrated that bisphenol A, a chemical commonly used in the production of plastics, functions as an endocrine disruptor, impacting organs like the thyroid, thymus, and pancreas. This is because it can attach to a variety of receptors connected to the endocrine system, impairing their functionality (Radwan et al., 2023). BPA is harmful to people, animals, and plants. According to some studies, BPA exposure leads to a variety of endocrinological, reproductive, and metabolic disorders in people. These disorders include diabetes, obesity, hypertension, heart disease, and cancers of the breast, ovaries, prostate, testicles, and endometrium (Siracusa et al., 2018). In response to these worries, several nations have outlawed its use in consumer goods, including infant bottles (Catenza et al., 2021). The number of elongated spermatids in the seminiferous tubules of pubertal ICR mice and the number of sperms in Holtzman rats decreased when BPA was administered to the animals (Siracusa et al., 2018).

Ampicillin is an antibiotic that causes an ecological imbalance in aquatic environments by poisoning aquatic life, changing plant growth, causing anomalies in the anatomy of many species, and fostering the growth of bacteria resistant to drugs (Okoye et al., 2022). When used excessively, paracetamol can have negative effects that include breast cancer cell proliferation. The usual explanation for this toxicity is reactive oxygen species, which can damage DNA and cause lipid peroxidation, protein denaturation, and other effects (Montaseri and Forbes, 2018). However, atrazine is an herbicide that is used in a variety of crops to suppress weeds. But this substance can also have an impact on other organisms, either directly or indirectly (Vieira et al., 2021). ATZ has also been shown to induce DNA methylation in the brain of carp and autophagy in the liver. It has also been shown to have an impact on human health through skin and respiratory contact, resulting in the development of breast and ovarian cancer as well as changes to the human vascular system (He et al., 2019).

11.6 WAYS OF MITIGATION AND THE ECONOMIC COST IN THE MANAGEMENT OF ECs

Scientists have adopted several mitigation technologies for emerging pollutants to lessen their effects on our environment. Three types of current mitigation techniques exist: chemical, biological, and physical methods. To mitigate the negative impacts of these pollutants, methods like adsorption, advanced oxidation processes (AOPs), and biological treatment have been investigated (Olatunde et al., 2020). Chemical oxidation is a technique used to transform pollutants into a harmless form. Chemical treatment generally refers to the use of chemicals in a series of reactions to facilitate the process of disinfecting wastewater (Ahmed et al., 2021). Physical treatment methods are, therefore, widely employed because they are easy to use and versatile. Adsorption and membrane technologies, such as micro-, ultra-, and nanofiltration as well as reverse osmosis, are used frequently in physical processes. Also, by using a range of microorganisms for biodegradation, biological treatment eliminates pollutants. Created wetlands, membrane bioreactors (MBRs), and activated sludge processes are the three primary nonconventional biological processes (Ahmed et al., 2021).

11.6.1 Enzyme Immobilization Technologies

The use of enzymes and their biocatalytic activity have been a strategy for removing pollutants in wastewater. However, these biocatalytic processes are usually limited by the lack of stability of the enzyme, the half-life time, and the feasibility of commercial application viability. Therefore, immobilization techniques have been implemented to improve its catalytic characteristics and reuse. Piao et al. (2019) used stabilized laccase in mesoporous silica to build a high-efficiency BPA treatment system in a fluidized bed reactor. Laccase is one of the most robust biocatalysts, with a wide range of uses in environmental processes such as detecting and treating chemical contaminants and colors, as well as pharmaceutical elimination. Silica particles are very attractive, as an immobilization matrix, because they have a well-ordered porous structure, high surface area, thick walls, and good hydrothermal resistance. The biotransformation of BPA, in the circulating fluidized bed reactor with immobilized laccases in silica, was very fast. 90% of BPA was transformed in eight hours (Piao et al., 2019).

Natural enzymes have been applied as an effective strategy to treat endocrine disruptors. For example, oxidases can catalyze estrogens to produce phenoxyl radical intermediates, leading to the subsequent formation of homo-oligomers with lower biotoxicity and bioavailability (Sun et al., 2020). However, their application is limited by the low stability and high-cost manufacture and storage of natural enzymes. Consequently, artificial nanozymes with simultaneous enzymatic and nanomaterial properties have been developed, being useful for environmental remediation purposes (Sun et al., 2020).

11.6.2 Biological Treatment

This mitigation strategy reduces wastewater through biodegradation and biosorption, and it is also widely used to treat newly emerging pollutants (He et al., 2019). Numerous researchers have examined how these substances biodegrade in a variety of systems, such as membrane bioreactors, sequential batch reactors, and constructed wetlands (Langbehn et al., 2021).

In general, a variety of factors, including the type of water matrix, the organisms' characteristics, the operating conditions, and the pollutants themselves, influence how quickly emerging pollutants biodegrade in biological processes (Vieira et al., 2021).

11.6.3 Advanced Oxidation

These are chemically treated organic compounds where strong oxidizing agents are added to carry out the oxidation of contaminants. These procedures are designed to break down the contaminants into more manageable products (Dave and Das, 2021). To assess the oxidation of various EDCs (E1, estriol [E3], E2, EE2, and BPA) in the presence of methanol in artificial solutions, Zong et al. (2021) examined five distinct structural phases of MnO2. The most reactive phase to the oxidation of E1 was the δ-MnO2, which was eliminated at pH 3 after 120 minutes. The overall performance of all structural phases of MnO2 was very poor (>45%) for the remaining EDCs.

Among the advanced oxidation techniques used to reduce concentrations of hormones and PPCPs in wastewater treatment in plants is ozone oxidation which has been identified as the most promising technique (Si et al., 2018). Ozonation is an advanced oxidation technology where ozone (O_3) is added to water, and due to its powerful oxidation capacity, the ozone produces reactive oxygen species that can degrade a wide range of compounds and microorganisms (Poddar and Sarkar, 2020). This technology has also been employed in several wastewater treatment in plants with promising results for the degradation of diverse contaminants, including some EDCs.

Avşar and Güzel (2020) conducted a study in which they optimized the ozonation conditions to remove E1 and E2 from synthetic solutions. The results of the experiments indicated that pH 6, ozone concentration of 4 mg L^{-1}, and residence time of 5 min were ideal. Approximately 90% of E1 and 95% of E2 were eliminated in these circumstances. The removal of gemfibrozil and ibuprofen from spiked treated sewage effluent was assessed using ozone and ozone/hydrogen peroxide treatments. By using O_3/H_2O_2 for both ozonation and AOP, the full removal of gemfibrozil and the 80% removal of ibuprofen were accomplished. One of the most promising methods for breaking down EDCs in wastewater treatment in plants is ozonation (Farzaneh et al., 2020).

11.6.4 Flocculation

The process of flocculation involves adding a chemical agent to a solution to facilitate the bonding between particles and produce larger aggregates that are easier to remove (Rodríguez-Hernandez et al., 2022). An additional separation technique, like filtration or decantation, must be used in conjunction with this one. A study using a polymeric cellulose-based flocculant was conducted to assess the removal of 4-nonylphenol from water with the coexistence of natural organic matter and suspended inorganic particles (Yang et al., 2016). The outcomes demonstrated that at a pH of 4, 35°C, and a flocculant dosage of 40 mg L^{-1}, removal was up to 79%. The removal of 4-nonylphenol benefited from the synergistic interaction between kaolin and natural organic matter.

A popular method for filtering impurities out of water is flocculation. Nevertheless, the majority of research focuses on eliminating ionic pollutants, and there isn't much evidence supporting their use in eliminating EDCs. Flocculation is not a good choice for a primary removal technology because the majority of studies on EDC removal that have been published have demonstrated negligible efficiencies when compared to other technologies like adsorption. To boost flocculation's effectiveness, more studies might be done on how to combine it with other removal techniques. In plants that currently treat wastewater using flocculation technology, this may be especially beneficial (Rodríguez-Hernandez et al., 2022).

11.6.5 Adsorption

This is one of the methods for removing impurities from water that is most frequently used and researched. This method is predicated on a contaminant's (adsorbate) adhesion to a solid material's (adsorbent) surface as a result of various physical or chemical interactions (Das and Ray, 2023). According to Wang et al. (2021), there are two main types of adsorption mechanisms: chemical adsorption, which is associated with the formation of chemical bonds, and physical adsorption, which is linked to ion exchange and van der Waals force.

Due to its ease of use, low cost, efficiency, and environmental friendliness, adsorption is a commonly used technique for the removal of different contaminants from water and wastewater (Benjelloun et al., 2021). Numerous factors, including the type and concentration of the adsorbent and adsorbate, the presence of additional pollutants, temperature, and experimental parameters like adsorbent surface, pH, and contact time, can affect this process (Vieira et al., 2021). Numerous investigations have been carried out regarding the adsorption of emerging pollutants by various substances. Carbon-based adsorbents have been used in some studies to remove a variety of pollutants. Zbair et al. (2018) investigated how activated carbon made from argan nut shells and activated by phosphoric acid could adsorb bisphenol A from water, eliminating 1,250 mg/g of BPA. The variables that were optimized were the pH, initial BPA concentration, adsorbent dosage, and 60 mg L^{-1} and 6.5, respectively. Adsorption tests showed that the adsorbent's high specific surface area (1,372 m^2/g) contributed to its increased efficiency. The obtained adsorption data showed a strong correlation with the Langmuir isotherm and the pseudo-second-order model.

The use of adsorption to remove newly emerging pollutants has been studied by numerous researchers. Since activated carbon is so effective at eliminating various kinds of pollutants, it is the best and most commonly used adsorbent in the world. Nevertheless, it is costly and challenging to regenerate, necessitating the hunt for substitute materials with comparable effectiveness. Adsorbent dosage, pollutant concentration, solution pH, contact time, temperature, adsorbent, adsorbate nature, and the presence of additional pollutants are just a few of the variables that can impact the adsorption process (Benjelloun et al., 2021).

11.6.6 Photocatalysis

This environmentally friendly method of eliminating organic pollutants from wastewater uses photocatalysts that can be activated by light irradiation. The process of photocatalysis involves the reaction of organic pollutants with strong oxidizing and reducing agents (h^+ and e^-) produced on the surface of photocatalysts by a light source (Paumo et al., 2021). Due to its many benefits in the degradation of pollutants, titanium dioxide (TiO_2) is the most commonly used photocatalyst. TiO_2 is not as good at absorbing solar radiation. Ahmed et al. (2021) reported that to improve its ability to absorb visible light, the majority of research focuses on altering it by doping it with metals (Ag^+, Fe^{3+}, and Co^{3+}) and nonmetals (N, S, F, C, B, and P). In addition, other oxides and perovskites (such as ZnO, WO_3, V_2O_5, $BiVO_4$, Ag_3VO_4, and $SrTiO_3$), bismuth oxyhalides (such as BiOCl, BiOBr, and BiOI), sulfides (such as CdS, ZnS, and MoS_2), and various composite materials have drawn the attention of researchers looking into photocatalytic materials for wastewater treatment applications (Antonopoulou et al., 2021).

11.6.7 Hybrid Process

The development of hybrid technologies, which use the various removal potentials of different processes to overcome the limitations of the removal of these compounds, is encouraged by the limited effectiveness of conventional treatment processes for the removal of many emerging pollutants (García et al., 2021). To effectively remove recalcitrant micropollutants, the hybrid process combines two or more treatment techniques (Grandclément et al., 2017).

11.6.8 Phytoremediation

Phytoremediation is a viable method for cleaning contaminated soils. This approach reduces pollutant toxicity in contaminated areas by utilizing plant interactions at the physical, biological, chemical, biochemical, and microbiological levels. Numerous techniques are used in phytoremediation, depending on the type and amount of pollutant (Nedjimi, 2021). The common techniques for eliminating contaminants such as heavy metals are extraction, sequestration, and transformation. Organic contaminants like oils and chloro-compounds can be broken down, immobilized, rhizoremediated, and evaporated by employing plants like willow or alfalfa (Awa and Hadibarata, 2020). Plants that act as phytoremediators have a variety of important characteristics, including their tap or fibrous root systems, penetration, toxicity levels, ability to withstand the harsh environmental conditions of the contaminants, annual growth, supervision, and—most importantly—the amount of time it takes for the environmental standards to be met. According to Chen et al. (2018), the plant must also be resistant to insects and diseases. Eliminating pollutants from the roots and shoots is a critical component of phytoremediation. Transpiration and partitioning are also necessary for the movement of nutrients and water (Awa and Hadibarata, 2020). This process can be changed by pollutants and the nature of plants. Most of the plants that are present at a polluted site can aid in phytoremediation.

11.7 CONCLUSION AND FUTURE RECOMMENDATION

In conclusion, managing the challenges posed by emerging pollutants in our environment requires a multifaceted approach that involves source control, treatment technologies, public awareness, and monitoring and research. Through a combination of these strategies, we can mitigate the impact of emerging pollutants on the environment and public health. Despite these strategies, there is still much to be done in terms of evaluating and addressing emerging pollutants on a global level. The state of the art of evaluating emerging pollutants globally is still evolving, with more studies being conducted and new pollutants being identified. Therefore, continuous monitoring and evaluation should be critical in developing effective strategies to manage emerging pollutants globally.

REFERENCES

Ahmed, S., Khan, F.S.A., Mubarak, N.M., Khalid, M., Tan, Y.H., Mazari, S.A., Karri, R.R. and Abdullah, E.C., 2021. Emerging pollutants and their removal using visible-light responsive photocatalysis–a comprehensive review. *Journal of Environmental Chemical Engineering* 9(6):106643.

Antonopoulou, M., Kosma, C., Albanis, T. and Konstantinou, I., 2021. An overview of homogeneous and heterogeneous photocatalysis applications for the removal of pharmaceutical compounds from real or synthetic hospital wastewaters under lab or pilot scale. *Science of the Total Environment* 765:144163.

Avşar Teymur, Y. and Güzel, F., 2020. Using of magnetized and non-magnetized tomato industrial processing solid waste in remediation of Reactive Blue 19 dye aqueous solution. *International Journal of Phytoremediation* 22(13):1420–1430.

Awa, S.H. and Hadibarata, T., 2020. Removal of heavy metals in contaminated soil by phytoremediation mechanism: A review. *Water, Air, and Soil Pollution* 231(2):47.

Bai, X., Lutz, A., Carroll, R., Keteles, K., Dahlin, K., Murphy, M. and Nguyen, D., 2018. Occurrence, distribution, and seasonality of emerging contaminants in urban watersheds. *Chemosphere* 200:133–142.

Bayabil, H.K., Teshome, F.T. and Li, Y.C., 2022. Emerging contaminants in soil and water. *Frontiers in Environmental Science* 10:873499.

Benjelloun, M., Miyah, Y., Evrendilek, G.A., Zerrouq, F. and Lairini, S., 2021. Recent advances in adsorption kinetic models: Their application to dye types. *Arabian Journal of Chemistry* 14(4):103031.

Catenza, C.J., Farooq, A., Shubear, N.S. and Donkor, K.K., 2021. A targeted review on fate, occurrence, risk and health implications of bisphenol analogues. *Chemosphere* 268:129273.

Chen, Y., Ding, Q., Chao, Y., Wei, X., Wang, S. and Qiu, R., 2018. Structural development and assembly patterns of the root-associated microbiomes during phytoremediation. *Science of the Total Environment*6 44:1591–1601.

Costa, E., Piazza, V., Lavorano, S., Faimali, M., Garaventa, F. and Gambardella, C., 2020. Trophic transfer of microplastics from copepods to jellyfish in the marine environment. *Frontiers in Environmental Science* 8:571732.

Das, S. and Ray, S., 2023. Adsorptive removal of aqueous pollutants on waste biomass-derived activated carbon: A mechanistic evaluation. *Materials Today: Proceedings* 77:156–162.

Dave, S. and Das, J., 2021. Technological model on advanced stages of oxidation of wastewater effluent from food industry. In *Advanced Oxidation Processes for Effluent Treatment Plants* (pp. 33–49). Elsevier.

Faheem, M. and Bhandari, R.K., 2021. Detrimental effects of bisphenol compounds on physiology and reproduction in fish: A literature review. *Environmental Toxicology and Pharmacology* 81:103497.

Fan, X., Wang, Z., Li, Y., Wang, H., Fan, W. and Dong, Z., 2021. Estimating the dietary exposure and risk of persistent organic pollutants in China: A national analysis. *Environmental Pollution* 288:117764.

Farzaneh, H., Loganathan, K., Saththasivam, J. and McKay, G., 2020. Ozone and ozone/hydrogen peroxide treatment to remove gemfibrozil and ibuprofen from treated sewage effluent: Factors influencing bromate formation. *Emerging Contaminants* 6:225–234.

Gallé, T., Bayerle, M., Pittois, D. and Huck, V., 2020. Allocating biocide sources and flow paths to surface waters using passive samplers and flood wave chemographs. *Water Research* 173:115533.

Gani, K.M., Hlongwa, N., Abunama, T., Kumari, S. and Bux, F., 2021. Emerging contaminants in South African water environment-a critical review of their occurrence, sources and ecotoxicological risks. *Chemosphere* 269:128737.

García, L., Leyva-Díaz, J.C., Díaz, E. and Ordóñez, S., 2021. A review of the adsorption-biological hybrid processes for the abatement of emerging pollutants: Removal efficiencies, physicochemical analysis, and economic evaluation. *Science of the Total Environment* 780:146554.

Gomes, N.O., Mendonça, C.D., Machado, S.A., Oliveira, O.N. and Raymundo-Pereira, P.A., 2021. Flexible and integrated dual carbon sensor for multiplexed detection of nonylphenol and paroxetine in tap water samples. *Microchimica Acta*188:1–10.

Grandclément, C., Seyssiecq, I., Piram, A., Wong-Wah-Chung, P., Vanot, G., Tiliacos, N., Roche, N. and Doumenq, P., 2017. From the conventional biological wastewater treatment to hybrid processes, the evaluation of organic micropollutant removal: A review. *Water Research* 111:297–317.

Han, X., Zuo, Y.T., Hu, Y., Zhang, J. and Zhou, M.X., 2018. Investigating the performance of three modified activated sludge processes treating municipal wastewater in organic pollutants removal and toxicity reduction. *Ecotoxicology and Environmental Safety* 148:729–737.

He, H., Liu, Y., You, S., Liu, J., Xiao, H. and Tu, Z., 2019. A review on recent treatment technology for herbicide atrazine in contaminated environment. *International Journal of Environmental Research and Public Health* 16(24):5129.

He, W., Hu, Z.H., Yuan, S., Zhong, W.H., Mei, Y.Z. and Dai, C.C., 2018. Bacterial bioreporter-based mercury and phenanthrene assessment in Yangtze River Delta Soils of China. *Journal of Environmental Quality* 47(3):562–570.

Huang, C., Zhao, J., Lu, R., Wang, J., Nugen, S.R., Chen, Y. and Wang, X., 2023. A phage-based magnetic relaxation switching biosensor using bioorthogonal reaction signal amplification for salmonella detection in foods. *Food Chemistry* 400:134035.

Hui, C.Y., Guo, Y., Gao, C.X., Li, H., Lin, Y.R., Yun, J.P., Chen, Y.T. and Yi, J., 2022. A tailored indigoidine-based whole-cell biosensor for detecting toxic cadmium in environmental water samples. *Environmental Technology & Innovation* 27:102511.

Jacob, R.S., de Souza Santos, L.V., d'Auriol, M., Lebron, Y.A.R., Moreira, V.R. and Lange, L.C., 2020. Diazepam, metformin, omeprazole and simvastatin: A full discussion of individual and mixture acute toxicity. *Ecotoxicology* 29(7):1062–1071.

Kassotis, C.D., Vandenberg, L.N., Demeneix, B.A., Porta, M., Slama, R. and Trasande, L., 2020. Endocrine-disrupting chemicals: Economic, regulatory, and policy implications. *Lancet Diabetes Endocrinol* 8(8):719–730. http://doi.org/10.1016/S2213-8587(20)30128-5.

Langbehn, R.K., Michels, C. and Soares, H.M., 2021. Antibiotics in wastewater: From its occurrence to the biological removal by environmentally conscious technologies. *Environmental Pollution* 275:116603.

Li, Z., Hou, J.T., Wang, S., Zhu, L., He, X. and Shen, J., 2022. Recent advances of luminescent sensors for iron and copper: Platforms, mechanisms, and bio-applications. *Coordination Chemistry Reviews* 469:214695.

Liu, Y. and Wang, T., 2020. Worsening urban ozone pollution in China from 2013 to 2017–Part 2: The effects of emission changes and implications for multi-pollutant control. *Atmospheric Chemistry and Physics* 20(11):6323–6337.

Liu, Y., Yang, X., Tan, J. and Li, M., 2023. Concentration prediction and spatial origin analysis of criteria air pollutants in Shanghai. *Environmental Pollution* 327:121535.

Mishra, R.K. and Mohanty, K., 2023. A review of the next-generation biochar production from waste biomass for material applications. *Science of the Total Environment* 6:74–95.

Montaseri, H. and Forbes, P.B., 2018. Analytical techniques for the determination of acetaminophen: A review. *TrAC Trends in Analytical Chemistry* 108:122–134.

Mpatani, F.M., Aryee, A.A., Kani, A.N., Guo, Q., Dovi, E., Qu, L., Li, Z. and Han, R., 2020. Uptake of micropollutant-bisphenol A, methylene blue and neutral red onto a novel bagasse-β-cyclodextrin polymer by adsorption process. *Chemosphere* 259:127439.

Nedjimi, B., 2021. Phytoremediation: A sustainable environmental technology for heavy metals decontamination. *SN Applied Sciences* 3(3):286.

Ng, B., Quinete, N., Maldonado, S., Lugo, K., Purrinos, J., Briceño, H. and Gardinali, P., 2021. Understanding the occurrence and distribution of emerging pollutants and endocrine disruptors in sensitive coastal South Florida Ecosystems. *Science of the Total Environment* 757:143720.

Nováková, Z., Novák, J., Kitanovski, Z., Kukučka, P., Smutná, M., Wietzoreck, M., Lammel, G. and Hilscherová, K., 2020. Toxic potentials of particulate and gaseous air pollutant mixtures and the role of PAHs and their derivatives. *Environment International* 139:105634.

Okoye, C.O., Nyaruaba, R., Ita, R.E., Okon, S.U., Addey, C.I., Ebido, C.C., Opabunmi, A.O., Okeke, E.S. and Chukwudozie, K.I., 2022. Antibiotic resistance in the aquatic environment: Analytical techniques and interactive impact of emerging contaminants. *Environmental Toxicology and Pharmacology* 96:103995.

Olatunde, O.C., Kuvarega, A.T. and Onwudiwe, D.C., 2020. Photo enhanced degradation of contaminants of emerging concern in waste water. *Emerging Contaminants* 6:283–302.

Paumo, H.K., Dalhatou, S., Katata-Seru, L.M., Kamdem, B.P., Tijani, J.O., Vishwanathan, V., Kane, A. and Bahadur, I., 2021. TiO_2 assisted photocatalysts for degradation of emerging organic pollutants in water and wastewater. *Journal of Molecular Liquids* 331:115458.

Peña-Guzmán, C., Ulloa-Sánchez, S., Mora, K., Helena-Bustos, R., Lopez-Barrera, E., Alvarez, J. and Rodriguez-Pinzón, M., 2019. Emerging pollutants in the urban water cycle in Latin America: A review of the current literature. *Journal of Environmental Management* 237:408–423.

Piao, M., Zou, D., Yang, Y., Ren, X., Qin, C. and Piao, Y., 2019. Multi-functional laccase immobilized hydrogel microparticles for efficient removal of bisphenol A. *Materials*1 2(5):704.

Pitter, G., Da Re, F., Canova, C., Barbieri, G., Zare Jeddi, M., Daprà, F., Manea, F., Zolin, R., Bettega, A.M., Stopazzolo, G. and Vittorii, S., 2020. Serum levels of perfluoroalkyl substances (PFAS) in adolescents and young adults exposed to contaminated drinking water in the Veneto region, Italy: A cross-sectional study based on a health surveillance program. *Environmental Health Perspectives* 128(2):027007.

Poddar, K. and Sarkar, A., 2020. Emerging treatment strategies of pharmaceutical pollutants: Reactive physiochemical and innocuous biological approaches. In *Removal of Toxic Pollutants Through Microbiological and Tertiary Treatment* (pp. 431–451). Elsevier.

Pokkiladathu, H., Farissi, S., Sakkarai, A. and Muthuchamy, M., 2022. Degradation of bisphenol A: A contaminant of emerging concern, using catalytic ozonation by activated carbon-impregnated nanocomposite-bimetallic catalyst. *Environmental Science Pollution Research* 4:1–14.

Radwan, E.K., Abdel Ghafar, H.H., Ibrahim, M.B.M. and Moursy, A.S., 2023. Recent trends in treatment technologies of emerging contaminants. *Environmental Quality Management* 32(3):7–25.

Rasheed, T., Bilal, M., Nabeel, F., Adeel, M. and Iqbal, H.M., 2019. Environmentally-related contaminants of high concern: Potential sources and analytical modalities for detection, quantification, and treatment. *Environment International* 122:52–66.

Rodríguez-Hernández, J.A., Araújo, R.G., López-Pacheco, I.Y., Rodas-Zuluaga, L.I., González-González, R.B., Parra-Arroyo, L., Sosa-Hernández, J.E., Melchor-Martínez, E.M., Martínez-Ruiz, M., Barceló, D. and Pastrana, L., 2022. Environmental persistence, detection, and mitigation of endocrine disrupting contaminants in wastewater treatment plants–a review with a focus on tertiary treatment technologies. *Environmental Science: Advances* 1:680–708.

Si, X., Hu, Z. and Huang, S., 2018. Combined process of ozone oxidation and ultrafiltration as an effective treatment technology for the removal of endocrine-disrupting chemicals. *Applied Sciences* 8(8):1240.

Siracusa, J.S., Yin, L., Measel, E., Liang, S. and Yu, X., 2018. Effects of bisphenol A and its analogs on reproductive health: A mini review. *Reproductive Toxicology* 79:96–123.

Srivastava, S., Kumar, A., Bauddh, K., Gautam, A.S. and Kumar, S., 2020. 21-day lockdown in India dramatically reduced air pollution indices in Lucknow and New Delhi, India. *Bulletin of Environmental Contamination and Toxicology* 105:9–17.

Sun, S., Jiang, T., Lin, Y., Song, J., Zheng, Y. and An, D., 2020. Characteristics of organic pollutants in source water and purification evaluations in drinking water treatment plants. *Science of the Total Environment* 733:139277.

Varsha, M., Kumar, P.S. and Rathi, B.S., 2022. A review on recent trends in the removal of emerging contaminants from aquatic environment using low-cost adsorbents. *Chemosphere* 287:132270.

Vieira, K.S., Neto, J.A.B., Crapez, M.A.C., Gaylarde, C., da Silva Pierri, B., Saldaña-Serrano, M., Bainy, A.C.D., Nogueira, D.J. and Fonseca, E.M., 2021. Occurrence of microplastics and heavy metals accumulation in native oysters Crassostrea Gasar in the Paranaguá estuarine system, Brazil. *Marine Pollution Bulletin* 166:112225.

Wang, G.H., Tsai, T.H., Kui, C.C., Cheng, C.Y., Huang, T.L. and Chung, Y.C., 2021. Analysis of bioavailable toluene by using recombinant luminescent bacterial biosensors with different promoters. *Journal of Biological Engineering* 15:1–9.

Yang, R., Li, H., Huang, M., Yang, H. and Li, A., 2016. A review on chitosan-based flocculants and their applications in water treatment. *Water Research* 95:59–89.

Zahmatkesh, S., Amesho, K.T. and Sillanpää, M., 2022. A critical review on diverse technologies for advanced wastewater treatment during SARS-CoV-2 pandemic: What do we know? *Journal of Hazardous Materials Advances* 7:1–22.

Zbair, M., Anfar, Z., Khallok, H., Ahsaine, H.A., Ezahri, M. and Elalem, N., 2018. Adsorption kinetics and surface modeling of aqueous methylene blue onto activated carbonaceous wood sawdust. *Fullerenes, Nanotubes and Carbon Nanostructures* 26(7):433–442.

Zhang, Y., Zhu, Y., Zeng, Z., Zeng, G., Xiao, R., Wang, Y., Hu, Y., Tang, L. and Feng, C., 2021. Sensors for the environmental pollutant detection: Are we already there? *Coordination Chemistry Reviews* 431:213681.

Zong, W., Guo, Z., Wu, M., Yi, X., Zhou, H., Jing, S., Zhan, J., Liu, L. and Liu, Y., 2021. Synergistic multiple active species driven fast estrone oxidation by δ-MnO2 in the existence of methanol. *Science of the Total Environment* 761:143201.

12 Legislative Framework in Addressing Emergent Pollutants and Ecological Impacts

Paul Atagamen Aidonojie and Osikemekha Anthony Anani

12.1 INTRODUCTION

The increasing complexity of modern industrial processes, technological innovations, and changing consumption patterns has led to the emergence of pollutants that pose novel challenges to environmental sustainability (Abbass et al., 2022). These substances, referred to as emergent pollutants, encompass a wide array of chemicals and contaminants whose impacts on ecosystems and human health are not fully understood at the time of their introduction (Anani et al., 2023a, 2023b).

Emergent pollutants challenge traditional regulatory paradigms due to their propensity for unexpected environmental persistence, bioaccumulation, and adverse effects on living organisms. The rapid pace of industrial innovation often outpaces regulatory processes, leading to a lag in identifying and mitigating the impacts of these pollutants. The ecological impacts of emergent pollutants are multifaceted and extend beyond immediate environmental degradation (Enuneku et al., 2021). These pollutants can disrupt ecosystems, harm wildlife, and potentially affect human health through bioaccumulation in the food chain (Hefft et al., 2022; Adama et al., 2023).

However, the dynamic nature of emergent pollutants necessitates a sophisticated and adaptive legislative framework to identify, monitor, regulate, and mitigate their ecological impacts. The legislative response to emergent pollutants is a crucial component of global environmental governance, reflecting society's commitment to preserving ecosystems, biodiversity, and the overall well-being of the planet (Alahi and Mukhopadhyay, 2017).

This framework serves as a guiding mechanism for enacting laws, regulations, and standards that address the unique characteristics and risks associated with emergent pollutants. Inherent in this legislative approach is the recognition of the intricate interplay between human activities, technological advancements, and the delicate balance of ecosystems (Aidonojie et al., 2023). As a result, the legislative framework needs to encompass a wide range of ecological considerations from the protection of biodiversity to the sustainable management of natural resources.

DOI: 10.1201/9781003362975-12

Furthermore, given the transboundary nature of emergent pollutants, international cooperation is paramount in addressing the challenges they pose. Collaborative efforts between nations can facilitate information exchange, research initiatives, and the development of shared regulatory standards (Andrady, 2015). Conventions such as the Stockholm Convention on Persistent Organic Pollutants and the Basel Convention on the Control of Transboundary Movements of Hazardous Wastes exemplify the global commitment to harmonizing legislative responses to emergent pollutants (Aidonojie, 2023).

However, despite the earlier example, there seems to be a threat to the environment arising from emergent pollutants. It is in this regard that this chapter explores the intricacies of legislative responses to emergent pollutants, shedding light on the multifaceted challenges and the imperative for adaptive, collaborative, and science-informed legislation in safeguarding the planet for present and future generations.

12.2 EMERGENT POLLUTANTS AND THEIR POSSIBLE IMPACT ON THE ENVIRONMENT AND HUMAN HEALTH

The issues from emergent pollutants have raised public concern for some decades now. Their routes and ecological and health impact from the least unicellular organisms to the highest animals (humans) in the ecosystem should be well studied and understood for proper assessment and future remediation if any.

Lei et al. (2015) looked at the human health impact of emergent pollutants. In their studies, they quantified that the general influence of emergent pollution on human health is not well-known. The most common health impacts are reproductive deformity and cancer. These are linked the contact with UV filters, pharmaceuticals (veterinary and humans alike), nanomaterials, additives from gasoline, and wastes from perfluorinated materials. From their reports, they opined that there was not too strong statistical evidence to establish the reproductive and cancer impacts of emergent pollutants. Hence, there have been no collaborated adverse influences between humans and emergent pollutants as well as their exposure limits at the period of this study investigations. However, there have been some preventive and monitoring methods, risk evaluation, and remediation on the safety of the emergent pollutants in the different environmental matrixes. In light of this, it should be noted that emergent pollutants removal from the environment and the body of humans require specialized methods and established remediation techniques.

In another study, Pereira et al. (2015) evaluated the human and ecological risks from emergent pollutants and their potential hazards. Medical, agricultural, and technological developments have been shown to cause the input of emergent pollution into the environment. Emergent pollutants show several chemical and chemical properties like long stay in the environment, poor degradation, and high biological accumulation in the environment. Personal care and pharmaceuticals are the major links to emergent pollutants in the ecosystem. Plasticizers, one of the emerging pollutants, have been noted to show high probable endocrine disruption and carcinogenic in living systems. Pesticides, one of the agrochemicals employed for farm purposes, have shown mutagenic and neurotoxic effects on various pests and the environment. The authors also stated that flame retardants used to prevent the propagation of flame during fire outburst episodes contain noxious properties that can elicit endocrine defects in biotas. Pereira et al. (2015) also reported that surfactants used as interfacial and surface properties in the fluid medium also contain chemicals that can affect changes in the endocrine system of many female species. Therefore, there is a need to monitor the possible hidden effects of emergent pollutants to provide proper legislation on their production and use to foster a green ecosystem.

Egbuna et al. (2021) in a study evaluated the incidences of emergent pollution in Nigeria and their possibility to cause health impacts on humans. The authors stated that emerging pollutants occur naturally or artificially in the environment. In nature, over 250 chemicals have been identified to fall under such name and into nine major categories like radiation or electromagnetic, radionuclides, mycotoxins, pesticides, organic compounds, volatile organic chemicals, polycyclic aromatic hydrocarbons, industrial inorganic and organic chemicals, pharmaceuticals, and personal care products. Other minor groups are particulate matter, microplastics, and microbes. These pollutants can be found in different ecological matrixes like sediments, soils, and surface and underground water. Egbuna et al. (2021) also reported that these pollutants behave in disguise as epigenetic and DNA programmers causing the alteration of the cells of living organisms, stress inducers, receptor blockers of β-adrenergic, and disruptors of the endocrine system in humans. This can pose a serious threat to the ecosystem and the biota therein.

12.3 INTERNATIONAL CONVENTIONS AND TREATIES ON EMERGENT POLLUTANTS; LEGAL ANALYSIS

Given the dangers posed by emergent pollutants within the global environment, the international community has sought the need to through a legal framework provide for the curtailment of the harmful activities that often result in the generation of the emergent pollutants. Some of these international treaties and conventions are briefly examined as follows.

12.3.1 Stockholm Convention on Persistent Organic Pollutants (POPs)

The Stockholm Convention on Persistent Organic Pollutants (POPs) of 2001 establishes a comprehensive framework for the regulation and control of substances posing risks to human health and the environment (Auta et al., 2017). The Stockholm Convention, in its preamble and Article 1, articulates its fundamental objectives and scope. It aims to protect human health and the environment from persistent organic pollutants. Article 1 specifies that the convention covers a range of substances listed in Annexes A, B, and C, including those that may be considered emergent pollutants due to their persistence and potential adverse effects. The convention's language aligns with the precautionary principle, emphasizing the need for anticipatory action to address emergent pollutants before they cause widespread harm.

The convention's approach to emergent pollutants is substantiated by Article 8, which outlines the procedure for adding new chemicals to Annexes A, B, or C. This provision allows for the identification and regulation of substances that exhibit emergent properties, ensuring that the convention remains adaptive to evolving scientific knowledge and emerging environmental threats (Hungerbuhler and Fiedler, 2016).

By designating specific pollutants for regulation, the Stockholm Convention, in Articles 5 and 6, establishes a structured framework for the control and reduction of releases of listed chemicals. This not only prevents the emergence of new threats but also addresses existing ones on a global scale. Furthermore, the convention's relevance is further captured by Article 10, which emphasizes the need for research, development, and exchange of information on the potential effects of the chemicals listed in the Annexes.

This provision ensures that parties remain informed about the properties and risks associated with listed substances, thereby facilitating a proactive approach to emergent pollutants (Aidonojie et al., 2023). In this way, the Stockholm Convention integrates the precautionary principle into the international regulatory landscape, acknowledging the dynamic nature of emergent pollutants and providing a mechanism for timely intervention (Aidonojie et al., 2022a, 2022b, 2022c).

Concerning the earlier example, it suffices to state that the Stockholm Convention's legal framework, as outlined in its relevant provisions, not only establishes the objectives and scope but also provides mechanisms for the identification, regulation, and control of emergent pollutants (Anani et al., 2022). Through its adaptive approach and emphasis on the precautionary principle, the convention stands as a crucial international instrument in safeguarding the global environment and human health from the adverse effects of persistent organic pollutants (Xiang et al., 2019).

12.3.2 Basel Convention on the Control of Transboundary Movements of Hazardous Wastes and Their Disposal

The Basel Convention on the Control of Transboundary Movements of Hazardous Wastes and Their Disposal, established in 1989, provides a legal framework for the management of hazardous wastes. The analysis later cites and examines pertinent provisions to elucidate the convention's approach to emergent pollutants (Ukhurebor and Aidonojie, 2021).

The Basel Convention's focus on hazardous waste management is explicitly outlined in its preamble and Articles 1 and 4. These provisions establish the convention's overarching objective to control transboundary movements of hazardous wastes and ensure their environmentally sound management. Article 1 defines hazardous wastes and, importantly, acknowledges the potential emergence of pollutants resulting from novel industrial processes or products within the ambit of the convention.

Furthermore, Article 4 underscores the comprehensive approach of the Basel Convention by emphasizing the reduction of hazardous waste generation and the promotion of environmentally sound management practices. By explicitly recognizing emergent pollutants that may arise from evolving industrial practices or products, the convention positions itself as a preemptive legal framework to address potential risks associated with the changing landscape of hazardous wastes (Aidonojie et al., 2022a, b).

The adaptability of the Basel Convention to emerging challenges is evident in its amendment process, as outlined in Article 17. This provision empowers parties to propose amendments, including the addition of new categories of hazardous wastes to Annexes I, II, or III. The dynamic legal feature allows the convention to stay relevant and responsive to emerging pollutants by incorporating new scientific knowledge and addressing evolving challenges. By regularly updating annexes to include new categories of hazardous wastes, the Basel Convention demonstrates its commitment to managing emergent pollutants effectively (Hungerbuhler and Fiedler, 2016; Aidonojie et al., 2021). This responsiveness fosters ongoing international collaboration, as parties collectively contribute to the convention's ability to address the changing landscape of hazardous materials. The amendment process thus becomes a crucial mechanism for adapting the legal framework to emerging challenges associated with emergent pollutants.

Concerning the earlier example, the Basel Convention, through its relevant provisions, not only establishes a comprehensive approach to hazardous waste management but also demonstrates a proactive stance in addressing emergent pollutants (Aidonojie et al., 2020). By explicitly recognizing the potential risks associated with evolving waste streams and incorporating an adaptable amendment process, the convention stands as a key international instrument for mitigating the environmental and health impacts of emergent pollutants on a global scale.

12.3.3 Rotterdam Convention

The Rotterdam Convention on the Prior Informed Consent Procedure for Certain Hazardous Chemicals and Pesticides in International Trade, established in 1998, plays a crucial role in addressing emergent pollutants. The pivotal role of the Rotterdam Convention in preventing the global proliferation of emergent pollutants is inferred from its primary focus on hazardous chemicals and pesticides in international trade.

Although not explicitly tailored for emergent pollutants, the convention's significance is highlighted in Articles 7 and 12. Article 7 establishes the prior informed consent procedure, requiring exporting parties to provide information to the importing country about the risks associated with specific chemicals. This procedure indirectly addresses emergent pollutants by fostering international cooperation and informed decision-making. Importing countries, through the receipt of comprehensive information, can make informed choices about whether to accept the import of specific chemicals, thereby preventing the inadvertent spread of substances with emergent properties. The convention's emphasis on transparency and informed consent is a vital component of preventing the global proliferation of emergent pollutants (Anderson et al., 2004).

Furthermore, Article 10 of the Rotterdam Convention emphasizes information exchange among parties, constituting an essential feature for addressing emergent pollutants. The provision encourages the sharing of information on the characteristics and risks of chemicals listed in Annex III, facilitating the early identification and evaluation of substances with emergent properties. This collaborative approach, reinforced by Article 11 on technical assistance, strengthens the ability of parties to collectively tackle potential hazards associated with emergent pollutants.

The commitment to shared information, as outlined in these provisions, underlines the convention's recognition of the importance of international cooperation in managing the complexities of rapidly evolving chemical landscapes (Bodansky et al., 2017). By providing a platform for the exchange of scientific knowledge and risk assessments, the Rotterdam Convention enables parties to stay abreast of emerging challenges and collectively respond to the intricacies associated with emergent pollutants.

The Rotterdam Convention, through its relevant provisions, establishes a robust framework for preventing the global proliferation of emergent pollutants. By incorporating a prior informed consent procedure and emphasizing information exchange mechanisms, the convention contributes significantly to international efforts to address emergent pollutants through cooperative and informed decision-making (Zhang et al., 2019).

Given the earlier review, the legal provisions within these international conventions collectively form a robust framework for addressing emergent pollutants. By integrating the precautionary principle, adapting to emerging challenges, and promoting information exchange, these legal instruments contribute significantly to global efforts in preventing, controlling, and managing the impact of emergent pollutants on human health and the environment.

12.4 REGIONAL INITIATIVES AND COLLABORATIONS ON EMERGENT POLLUTANTS

However, it must be noted that the issues of emergent pollutants have also received regional attention through regional laws and regulations. In this regard, some of these regional bodies' regulations as they concern emergent pollutants are considered as follows.

12.4.1 European Union Regulations on Emergent Pollutants

The European Union (EU) has long been at the forefront of environmental stewardship, and its regulatory framework reflects a robust commitment to addressing emergent pollutants. These regulatory frameworks consist of two key instruments: the Registration, Evaluation, Authorization, and Restriction of Chemicals (REACH) Regulation and the Circular Economy Framework that form the bedrock of the EU's strategy for managing emergent pollutants (Ijaiya et al., 2018).

The REACH Regulation, enacted under Regulation (EC) No 1907/2006, stands as a pivotal instrument in the EU's endeavor to manage emergent pollutants. Its overarching objective, articulated in Article 1, is to ensure a high level of protection for human health and the environment. The relevance of REACH to emergent pollutants is particularly provided for in Article 57, which designates certain substances as substances of very high concern (SVHC) based on their hazardous properties.

The dynamic nature of emergent pollutants, characterized by evolving risks and properties, finds resonance in the identification of SVHCs. By designating substances as SVHCs, REACH triggers a cascade of stricter control measures, exemplifying the EU's commitment to proactive chemical management (Aidonojie et al., 2022a, b, c). The regulation not only mandates the registration and evaluation of chemicals but also authorizes and restricts certain substances, creating a comprehensive framework that addresses emergent pollutants from their inception through to their use and disposal. The proactive approach of REACH is instrumental in staying ahead of potential risks posed by emergent pollutants. The regulation, through its extensive reach, exemplifies the EU's dedication to continuously adapting and refining its chemical management strategy, ensuring that emerging threats are met with decisive regulatory measures.

Furthermore, it suffices to state that in tandem with REACH, the EU employs the Circular Economy Framework to further its efforts in managing emergent pollutants. While not explicitly crafted for emergent pollutants, the Circular Economy Action Plan is a pivotal element in the EU's broader environmental agenda. The plan seeks to promote sustainability and reduce the environmental impact of products by prioritizing waste prevention, reuse, and efficient resource utilization.

In this regard, the Waste Framework Directive (2008/98/EC), a cornerstone of the Circular Economy Framework, places a strong emphasis on waste prevention and eco-design. Although emergent pollutants may not be explicitly named, the circular economy approach aligns with the overarching goal of minimizing the impact of emerging pollutants throughout the product life cycle. By encouraging eco-design principles and promoting recycling initiatives, the EU fosters sustainable practices that indirectly contribute to managing the environmental footprint of emergent pollutants.

The Circular Economy Framework, with its emphasis on sustainable practices and efficient resource use, represents a holistic strategy. It recognizes that a circular economy is not only economically beneficial but also ecologically imperative, particularly in the context of emergent pollutants where a life cycle perspective is crucial for effective management.

Concerning the earlier example, it can be said that the EU's regulatory approach to emergent pollutants, encapsulated in REACH and the Circular Economy Framework, demonstrates a comprehensive and proactive strategy. By incorporating emergent pollutants into the REACH framework and promoting sustainable practices through the Circular Economy Action Plan, the EU exemplifies a multifaceted approach aimed at protecting human health and the environment from the potential risks posed by evolving chemical substances.

This commitment is not just regulatory but also reflective of a broader vision for a sustainable and resilient future. The EU's holistic approach recognizes the interconnectedness of environmental, economic, and social factors in managing emergent pollutants, ensuring that regulatory measures are aligned with broader sustainability goals. As the global community faces increasingly complex challenges associated with emergent pollutants, the EU's regulatory framework serves as a beacon of proactive and adaptive environmental governance.

12.4.2 African Union Environmental Policies on Emergent Pollutants

The African Union (AU), in its pursuit of sustainable development and environmental conservation, has implemented policies that indirectly address emergent pollutants. The major regulatory framework of AU's environmental strategy includes the African Convention on the Conservation of Nature and Natural Resources and regional collaboration initiatives. This regulatory framework will be examined to shed light on how the continent is working towards preventing and managing emergent pollutants.

The African Convention on the Conservation of Nature and Natural Resources, adopted in 1968, serves as a foundational legal instrument for environmental conservation across the African continent. Although this convention doesn't explicitly target emergent pollutants, its relevance is deeply rooted in Articles 2 and 3.

These articles emphasize the sustainable use of natural resources and the conservation of ecosystems. Article 2 stipulates the imperative of utilizing natural resources in a manner that ensures their sustainability. By promoting responsible resource management, the convention indirectly contributes to the prevention of emergent pollutants that may arise from the exploitation and depletion of natural resources (Anderson et al., 2004).

Furthermore, Article 3 focuses on the conservation of ecosystems, recognizing their intrinsic value. Preserving ecosystems is crucial in averting ecological disruptions that could lead to the emergence of pollutants with adverse effects on both the environment and human health (Anani et al., 2023a, b). The African Convention thus provides a solid framework for addressing emergent pollutants indirectly, aligning with the AU's commitment to sustainable and responsible environmental practices. By emphasizing the prudent use of resources and the conservation of ecosystems, the convention lays the groundwork for a harmonious coexistence between development and environmental preservation (Ma et al., 2018).

Also, the AU's commitment to combating emergent pollutants is manifest in its regional collaboration efforts, epitomized by initiatives such as the African Ministerial Conference on the Environment (AMCEN). Although not bound by a specific legal instrument, these collaborative platforms serve as critical arenas for member states to collectively address emerging environmental challenges.

AMCEN, established in 1985, exemplifies the AU's dedication to fostering cooperation in the realm of environmental conservation. Through this platform, member states share knowledge, exchange experiences, and collectively strategize on addressing common environmental challenges. While not explicitly designed to combat emergent pollutants, the collaborative efforts within AMCEN indirectly contribute to the prevention and management of emergent pollutants by creating a space for the exchange of best practices and the development of collective strategies (Xiang et al., 2019).

The AU's broader vision, encapsulated in Agenda 2063, also plays a pivotal role in addressing emergent pollutants. Although expansive in scope, Agenda 2063 envisions a sustainable and resilient Africa. This overarching vision indirectly contributes to the prevention and management of emergent pollutants by emphasizing the importance of sustainable development and environmental conservation (Bodansky et al., 2017). The continent's resilience is intricately linked to its ability to manage emerging environmental challenges, including the potential risks posed by emergent pollutants.

The AU's environmental policies on emergent pollutants demonstrate a commitment to holistic and sustainable practices. While the African Convention on the Conservation of Nature and Natural Resources sets the foundation for responsible resource management and ecosystem conservation, regional collaboration initiatives such as AMCEN provide a dynamic platform for member states to collectively tackle emerging environmental challenges, including emergent pollutants (Ukhurebor and Aidonojie, 2021).

The AU's broader vision, as outlined in Agenda 2063, reinforces the importance of sustainability and resilience, indirectly contributing to the prevention and management of emergent pollutants across the continent. As the AU continues to evolve its environmental policies, the emphasis on collaboration and sustainability positions Africa to effectively address the complexities associated with emergent pollutants in the years to come (Zhang et al., 2019).

In furtherance of the earlier example, it suffices to state that regional initiatives within the European Union and the African Union demonstrate proactive approaches to emergent pollutants. Through specific regulations like REACH in the EU and overarching frameworks such as the African Convention, these regional entities integrate emergent pollutant considerations into their broader environmental agendas, showcasing a commitment to sustainable practices and cooperative efforts for environmental protection.

12.5 CHALLENGES AND GAPS IN THE LEGISLATIVE FRAMEWORK ON EMERGENT POLLUTANTS

The management of emergent pollutants poses a significant challenge to global environmental governance, necessitating a robust legislative framework. This examination focuses on the challenges and gaps inherent in the current legislative approach to emergent pollutants, shedding light on the complexities that legislators face in addressing these dynamic and evolving threats.

12.5.1 Rapid Pace of Innovation and Technological Advancement

One of the foremost challenges in the legislative framework for emergent pollutants is the rapid pace of innovation and technological advancement. The emergence of new chemicals and pollutants from evolving industrial processes and novel products outpaces the ability of regulatory bodies to identify, evaluate, and legislate against them effectively (Xiang et al., 2019). Traditional legislative approaches, often characterized by lengthy processes, struggle to keep up with the speed at which new pollutants enter the environment.

This lag poses a significant risk, as emergent pollutants may cause harm before regulatory measures can be implemented. The dynamic nature of industries and the constant evolution of chemical compounds necessitate legislative frameworks that are agile and adaptive. Establishing mechanisms to expedite the identification and assessment of emergent pollutants is crucial to closing this gap in the legislative response (Bodansky et al., 2017). Additionally, fostering collaboration between regulatory bodies, research institutions, and industries can enhance the timely identification of emerging threats and facilitate more agile legislative responses.

12.5.2 Lack of Specificity in Regulatory Definitions

Another notable challenge lies in the lack of specificity in regulatory definitions related to emergent pollutants. Traditional regulatory frameworks often use broad categories or generic terms, making it challenging to precisely capture the diverse characteristics of emergent pollutants. The absence of clear and specific definitions hampers the identification, monitoring, and regulation of these pollutants, contributing to gaps in enforcement and compliance (Anderson et al., 2004).

Addressing this challenge requires a shift towards more nuanced and specific definitions within legislative frameworks. Establishing criteria that reflect the unique properties of emergent pollutants, such as persistence, bioaccumulation, and toxicity, can enhance regulatory precision. Furthermore, continuous collaboration between scientific communities and legislative bodies is essential to refining definitions in response to evolving scientific understanding and emerging challenges.

12.5.3 Global Nature of Emergent Pollutants

Emergent pollutants often transcend national borders, requiring international cooperation and a harmonized legislative response. The global nature of emergent pollutants introduces challenges related to jurisdictional limitations and disparities in legislative frameworks among countries. Inconsistencies in regulations may create loopholes that allow pollutants to migrate across regions, impacting ecosystems and human health on a global scale (Barboza et al., 2019).

To address this challenge, there is a need for enhanced international collaboration and the development of standardized frameworks. Existing international conventions, such as the Stockholm Convention and the Basel Convention, provide a foundation for global cooperation, but efforts must be intensified to ensure that legislative measures are consistent, comprehensive, and universally applied. Strengthening the enforcement mechanisms of international agreements and fostering information exchange between nations are vital steps in mitigating the global challenges associated with emergent pollutants.

12.5.4 Limited Integration of Emerging Technologies

Legislative frameworks often lag in integrating emerging technologies that can play a pivotal role in addressing emergent pollutants. Technologies such as artificial intelligence, advanced sensors, and monitoring systems offer opportunities for more proactive identification and management of pollutants (Ukhurebor and Aidonojie, 2021).

However, existing legislation may not adequately account for these technologies or provide a regulatory framework to govern their use in addressing emergent pollutants. Bridging this gap requires a proactive approach by legislators to integrate emerging technologies into legislative frameworks. Establishing regulatory sandboxes or pilot programs that allow the controlled testing of new technologies can facilitate their integration. Furthermore, legislative bodies need to engage with the scientific and technological communities to stay informed about emerging innovations and ensure that regulations remain relevant and effective.

12.5.5 Limited Public Awareness and Participation

Legislative frameworks can encounter challenges in addressing emergent pollutants due to limited public awareness and participation. Public engagement is crucial for the success of environmental regulations, as it ensures that the concerns and insights of affected communities are considered in the legislative process. However, emergent pollutants, by their nature, may lack public awareness, making it difficult to garner public support and participation in the development and implementation of legislative measures.

12.5.6 Enforcement and Compliance Issues

Enforcing and ensuring compliance with legislative measures related to emergent pollutants presents another significant challenge. The complexities associated with these pollutants, including varying sources, dispersion pathways, and potential delayed impacts, make it challenging for regulatory authorities to monitor and enforce compliance effectively. Industries, particularly those engaged in innovative technologies or processes, may face difficulties in aligning their practices with evolving regulations, leading to gaps in enforcement (Xiang et al., 2019). To overcome enforcement challenges, legislative frameworks should emphasize the development of robust monitoring and reporting systems.

Implementation of stringent penalties for non-compliance and the establishment of incentives for adherence to regulations can act as deterrents. Regular audits, inspections, and the use of advanced technologies for real-time monitoring can enhance the enforcement capacity of regulatory bodies. Moreover, fostering a culture of corporate responsibility and collaboration between industries and regulators can contribute to a more effective enforcement ecosystem.

12.5.7 Need for Comprehensive Monitoring and Reporting Mechanisms

The need for comprehensive monitoring and reporting mechanisms is a crucial aspect of addressing emergent pollutants but often presents a notable gap in legislative frameworks. Monitoring emergent pollutants requires sophisticated technologies and methodologies capable of detecting substances with diverse properties and potential ecological impacts. Additionally, establishing effective reporting mechanisms is essential for collecting timely data on the presence and characteristics of emergent pollutants.

To bridge this gap, legislative bodies should invest in the development and deployment of state-of-the-art monitoring technologies. This includes sensor networks, remote sensing technologies, and advanced analytical methods capable of detecting and characterizing emergent pollutants in various environmental matrices. Furthermore, the establishment of standardized reporting requirements and protocols for industries, research institutions, and the public can contribute to a more comprehensive understanding of the presence and behavior of emergent pollutants.

Overcoming this challenge requires proactive efforts to raise public awareness about emergent pollutants and their potential impacts. Legislators and regulatory bodies should prioritize transparent communication, public education campaigns, and mechanisms for public input in the legislative process. Empowering communities to participate in monitoring and reporting emergent pollutants can enhance the effectiveness of legislative frameworks and contribute to the overall success of environmental protection efforts.

12.6 CONCLUSION

The legislative framework on emergent pollutants faces multifaceted challenges and gaps that necessitate careful consideration and innovative solutions. Addressing the rapid pace of innovation, refining regulatory definitions, fostering global cooperation, integrating emerging technologies, and enhancing public awareness are critical steps in fortifying the legislative response to emergent pollutants.

As we navigate the complexities of environmental governance in the face of evolving threats, an adaptive, collaborative, and technology-informed legislative approach is paramount to effectively addressing the challenges posed by emergent pollutants on a global scale.

12.7 RECOMMENDATION

Concerning the earlier example, the following are, therefore, recommended as follows:

1. Adaptive and responsive legislative framework as the catalyst.
2. Addressing emergent pollutants requires a collaborative and internationally coordinated effort. Collaboration between lawmakers, scientists, and environmental agencies must ensure that legislation is grounded in the latest scientific findings, fostering a holistic and evidence-based approach to environmental protection.
3. Public engagement and awareness.
4. There is a need for effective legislation that encourages sustainable practices, discourages environmentally harmful activities, and encourages industries to adopt circular economy principles.

REFERENCES

Adama, K.K., Ikhazuangbe, P.M.O., Onyeachu, I.B., Aliu, D. and Anani, O.A. 2023. Development of polyacrylamide from waste agricultural by-product for enhanced oil recovery process in Nigerian oil industry: Conceptual review. *Nigerian Research Journal of Engineering and Environmental Sciences* 8(2): 370–382.

Abbass, K., Qasim, M. Z., Song, H., et al. A Review of the Global Climate Change Impacts, Adaptation, and Sustainable Mitigation Measures. *Environmental Science and Pollution Research* 29, 42539–42559 (2022). https://doi.org/10.1007/s11356-022-19718-6.

Aidonojie, P. A. Environmental Hazard: The Legal Issues Concerning Environmental Justice in Nigeria. *Journal of Human Rights, Culture and Legal System* 3(1), 17–32 (2023). https://doi.org/10.53955/jhcls.v3i1.60

Aidonojie, P. A., Agbale, O. P., & Ikubanni, O. O. Analysing the Land Use Act, the Grazing Reserves Act and the Proposed Fulani Cattle Colonies (RUGA Settlement) and NLTP. *Nnamdi Azikiwe University Journal of International Law and Jurisprudence* 12(1), 138–148 (2021).

Aidonojie, P. A., Anani, O. A., Agbale, O. P., Olomukoro, J. O., & Adetunji, O. C. Environmental Law in Nigeria: A Review on Its Antecedence, Application, Judicial Unfairness and Prospects. *Archive of Science & Technology* 1(2), 211–221 (2020).

Aidonojie, P. A., Idahosa, M. E., Agbale, O. P., & Oyedeji, A. I. The Environmental Conservation, and Ethical Issues Concerning Herbal Products in Nigeria. *Journal of Environmental Science and Economics* 1(3), 26–32 (2022a). https://doi.org/10.56556/jescae.v1i3.124

Aidonojie, P. A., Ikubanni, O. O., Oyedeji, A. I., & Oyebade, A. A. The Legal Challenges and Effect concerning the Environmental Security in Nigeria: A Lesson from International Perspective. *Journal of Commercial and Property Law* 9(1), 110–120 (2022b).

Aidonojie, P. A., Okuonghae, N., Moses-oke, R. O., & Majekodunmi, T. A. A Facile Review on the Legal Issues and Challenges Concerning the Conservation and Preservation of Biodiversity. *Global Sustainability Research* 2(2), 34–46 (2023).

Aidonojie, P. A., Ukhurebor, K. E., Masajuwa, F., Imoisi, S. E., Edetalehn, O. I., & Nwazi, J. Legal Implications of Nanobiosensors Concerning Environmental Monitoring. In: Singh, R. P., Ukhurebor, K. E., Singh, J., Adetunji, C. O., & Singh, K. R. (eds) *Nanobiosensors for Environmental Monitoring.* Springer, Cham (2022c). https://doi.org/10.1007/978-3-031-16106-3_21

Aidonojie, P. A., Ukhurebor, K. E., Oaihimire, I. E., Ngonso, B. F., Egielewa, P. E., Akinsehinde, P. O., Kusuma, H. S., & Darmokoesoemo, H. Bioenergy Revamping and Complimenting the Global Environmental Legal Framework on the Reduction of Waste Materials: A Facile Review. *Heliyon* 9(1) (2023). https://doi.org/10.1016/j.heliyon.2023.e12860

Alahi, M. E., & Mukhopadhyay, S. C. Detection methodologies for pathogen and toxins: A review. *Sensors* 17(8), 1–8 (2017).

Anani, O. A., Abel, I., Osayomwanbo, O., Olisaka, F. N., Aidonojie, P. A., Olatunji, E. O., & Habib, A. I. Application of Microorganisms as Biofactories to Produce Biogenic Nanoparticles for Environmental Cleanup: Currents Advances and Challenges. *Current Nanoscience*19(6), 770–782 (2023a).

Anani, O. A., Aidonojie, P. A., & Olomukoro, J. O. *Environmental Principles and Ethics: Current Challenges in the Field of Bioscience and Law, Ethics, Media, Theology and Development in Africa: A Festchrift in Honour of Msgr. Prof. Dr. Obiora Francis Ike.* Global.net Co-Publication & Others, Geneva, Switzerland, 142–158 (2022).

Anani, O. A., Shah, M. P., Aidonojie, P. A., & Enuneku, A. A. Bio-Nano Filtration as an Abatement Technique Used in the Management and Treatment of Impurities in Industrial Wastewater. *Bio-Nano Filtration in Industrial Effluent Treatment*, 171–182 (2023b).

Anderson, P. K., Cunningham, A. A., Patel, N. G., Morales, F. J., Epstein, P. R., & Daszak, P. Emerging Infectious Diseases of Plants: Pathogen Pollution, Climate Change and Agrotechnology Drivers. *Trends in Ecology & Evolution* 19(10), 535–544 (2004).

Andrady, A. L. *Persistence of Plastic Litter in the Oceans. Marine Anthropogenic Litter.* Springer Nature, Berlin, 57–72 (2015).

Auta, H. S., Emenike, C. U., & Fauziah, S. H. Distribution and Importance of Microplastics in the Marine Environment: A Review of the Sources, Fate, Effects, and Potential Solutions. *Environmental International* 102, 165–176 (2017).

Barboza, L. G. A., Cózar, A., & Gimenez, B. C. *Macroplastics Pollution in the Marine Environment. World Seas: An Environmental Evaluation*. Elsevier, Amsterdam, 305–328 (2019).

Bodansky, D., Brunnee, J., & Rajamani, L. *International Climate Change Law.* Oxford University Press, Oxford (2017).

Egbuna, C., Amadi, C. N., Patrick-Iwuanyanwu, K. C., Ezzat, S. M., Awuchi, C. G., Ugonwa, P. O., & Orisakwe, O. E. Emerging Pollutants in Nigeria: A Systematic Review. *Environmental Toxicology and Pharmacology* 85: 103638 (2021). http://doi.org/10.1016/j.etap.2021.103638.

Enuneku, A., Anani, O. A., Job, O., Kubeyinje, B. F., Ogbomida, E. T., Asemota, C. O., Okpara, B., Imoobe, T., Ezemonye, L. I., Adetunji, C. O., & Hefft D. O. Mapping Soil Susceptibility to Crude oil Pollution in the Region of Delta, South-South Nigeria: A Proportional Study of Evironmetrics, Health, Ecological Risks, and Geospatial Evaluation. *Scientific African*, e01012 (2021). https://doi.org/10.1016/j.sciaf.2021.e01012.

Hefft, D. I., Anani, O. A., Aigbodion, F., Osadagbonyi, C., Adetunji, C. O., Ejomah, A., Osariyekemwen, C., & Enuneku, A. Ex-situ Studies on Macrotermes Bellicosus as a Potential Bioremediation Tool of Polluted Dump Soil Sites for Sub Saharan Africa. *Soil and Sediment Contamination: An International Journal* (2022). http://doi.org/10.1080/15320383.2021.2017402.

Hungerbuhler, K., & Fiedler, H. Ten Years After Entry into Force of the Stockholm Convention: What Do Air Monitoring Data Tell About Its Effectiveness? *Environment Pollution* 217, 149–158 (2016). https://doi. org/10.1016/j.envpol.2016.01.090.

Ijaiya, H., Wardah, I. A., & Wuraola, O. T. Re-Examining Hazardous Waste in Nigeria: Practice Possibilities within the United Nations System. *African Journal of International and Comparative Law* 26(2), 264–282 (2018).

Lei, M., Zhang, L., Lei, J., Zong, L., Li, J., Wu, Z., & Wang, Z. Overview of Emerging Contaminants and Associated Human Health Effects. *Biomed Research International.* 404796 (2015). http://doi.org/10.1155/2015/404796.

Ma, F., Zhang, C., & Li, C. Development of Quantum Dot-Based Biosensors: Principles and Applications. *Journal of Material Chemistry B* 6(39): 6173–6190 (2018).

Pereira, L. C., de Souza, A. O., Franco Bernardes, M. F., Pazin, M., Tasso, M. J., Pereira, P. H., & Dorta, D. J. A Perspective on the Potential Risks of Emerging Contaminants to Human and Environmental Health. *Environment Science Pollution Research Internation* 22(18), 13800–13823 (2015). http://doi.org/10.1007/s11356-015-4896-6.

Ukhurebor, K. E., & Aidonojie, P. A. The Influence of Climate Change on Food Innovation Technology: Review on Topical Developments and Legal Framework. *Agriculture & Food Security* 10(1), 1–14 (2021). https://doi.org/10.1186/s40066-021-00327-4

Xiang, L., Bo, W., Yuhua, C., Shuang, Z., Hang, W., Xiao, F., Junwen, Z., & Xiaojie, M. Water Contaminant Elimination Based on Metal–Organic Frameworks and Perspective on Their Industrial Applications. *ACS Substantiable Chemistry and Engineering* 7(5), 4548–4563 (2019). https://doi.org/10.1021/acssuschemeng.8b05751

Zhang, H., West, D., Shi, H., Ma, Y., Adams, C., & Eichholz, T. Simultaneous Determination of Selected Trace Contaminants in Drinking Water Using Solid-Phase Extraction-High Performance Liquid Chromatography-Tandem Mass Spectrometry. *Water Air Soil Pollution* 23, 2840 (2019).

Index

A

B

C

D

E

F

R

S

T

U

V

W

Z

Printed in the United States
by Baker & Taylor Publisher Services